Multi Platinum Pro Tools

Advanced editing, pocketing and autotuning techniques

Multi Platinum
Pro Tools

Advanced editing, pocketing and autotuning techniques

Nathan Adam

And

Brady Barnett

AMSTERDAM • BOSTON • HEIDELBERG • LONDON • NEW YORK • OXFORD
PARIS • SAN DIEGO • SAN FRANCISCO • SINGAPORE • SYDNEY • TOKYO
Focal Press is an imprint of Elsevier

ELSEVIER

Focal
Press

Focal Press is an imprint of Elsevier
Linacre House, Jordan Hill, Oxford OX2 8DP, UK
30 Corporate Drive, Suite 400, Burlington MA 01803, USA

First edition 2006

British Library Cataloguing in Publication Data
A catalogue record for this book is available from the British Library

Library of Congress Cataloging-in-Publication Data
A catalog record for this book is available from the Library of Congress

ISBN-13: 978-0-24-052023-0
ISBN-10: 0-24-052023-8

For information on all Focal Press publications
visit our website at www.books.elsevier.com

Typeset by Charon Tec Ltd, Chennai, India
www.charontec.com
Printed and bound in the Canada

Working together to grow libraries in developing countries

www.elsevier.com | www.bookaid.org | www.sabre.org

ELSEVIER   BOOK AID International   Sabre Foundation

Contents

About the authors   xi

Chapter 1   **Pocketing, editing, and tuning: The what, whys, and hows of major-label Pro Tools editing**   **1**

Welcome: How this project came to be   1

The times they are a changin'   2

A bit of history: The progression of performance creation using studio technology   4

The 80/20 rule, and how it applies to you   4

Why do we need another Pro Tools book?   5

Definitions: A familiar-feeling guide to some unfamiliar terms   6

Using the accompanying DVD-ROM   10

Chapter 2   **Adjusting your seat-belt: Setting up Pro Tools preferences for a streamlined edit session**   **13**

How to approach a session: Doing things the platinum record way   13

Configuring your session, choosing your settings, keeping your sanity   14

Pro Tools preferences and how they affect your session flow   15

Mix window, Edit window, or both? Why use two when you only need one?   42

Link Timeline and Edit Selection   45

Utilizing the Show/Hide button   46

Slip, Spot, Shuffle, Grid: A brief look at the major

editing modes and tools 48

Slip mode 49

Shuffle mode 53

Grid mode 55

Zooming 58

Zoom in/out 58

Shortcuts 65

The Smart tool 66

Chapter 3 **Beginning the pocket: Building a song**
from the drums up **69**

Opening the session 69

Importing tracks 71

Activating the track 74

Taking stock of the session 75

What to view and how to view it: Cleaning up the

Edit window 75

Finding a visual guide track 79

Editing within a drum group 82

Track-naming conventions 83

What to do when there is no kick or snare 88

Excess noise cleanup 88

Pocketing our first note 92

Moving on: A drum pocketing system 96

What to do when you have no transient to pocket to 99

Double transients in drum pocketing and how to

deal with them 102

Using time compression/expansion to fix a drum edit 104

The difference between Serato and Digi TCE tools 106

Proper use of time stretching when fixing drums 110

Adjusting the bounds to fit your fades 113

The key to good drum editing 116

Using pre-roll to check your drum pocket 116

Pocket it or leave it alone? A rule of thumb 117

Using Spot mode to bail out when you lose your

editing perspective 119

	Tab to Transients: The good and bad	121
	How much 'feel' to leave in the track	122
	Wrapping up the drums	124
Chapter 4	**Using Beat Detective to save time, money, and headaches**	**131**
	Importing the percussion tracks	132
	The Audio Media Options	133
	Setting up your percussion Edit window	135
	Pocketing the brush track	136
	Beat Detective	137
	Setting up for a good beat detection	138
	Launching and using Beat Detective to pocket the brush	140
	Keeping your frame of reference: Did you improve the track?	148
	Edit smoothing and filling gaps: The right choice	148
	Fitting it in the mix: Keeping your time priorities straight	151
	Special cases in Beat Detective and how to address them	152
	The Trigger Pad option	153
	A quick Beat Detective recap	157
	Fixing clashes between Beat Detective and the master drum take	159
	Using batch fades with Beat Detective	167
	Using the Pre-splice option to avoid double transients	169
	The lazy way out: Avoiding Beat Detective with Copy and Paste	170
Chapter 5	**Getting the bass player on time – and not just to rehearsal**	**173**
	Bass guitar pocketing: The setup	173
	Slip mode or Grid? Using your ears instead of your eyes	175
	The Show/Hide bin: Focusing your edit	176
	Utilizing the Zoom Waveform – grow transients, grow	178

A look at the pocket: Where does the bass wave
    begin, and where should it end?   180

Separate, Fade, and Nudge: A simple bass pocket   182

Two schools of thought for pocketing bass   184

Run, Spot, run: Spot mode comes to the rescue
    when nothing else will   185

A closer look at the bass   186

Pre-rolls, post-rolls, and solos: Repeat that five
    times fast   188

Double trouble, double bass transients, and how to
    fix them   191

Editing without crossfades: The 'nudge at the
    sample level' routine   195

Moving the regions, filling in the holes   197

Overcoming the beast of fret noise   199

Time compression and expansion: Using the
    TCE tool to fix bass gaps   200

When nothing else works – try Copy and Paste   203

Using your eyes *and* your ears: A wrap-up   205

Chapter 6   **Locking up the acoustic tracks**   **207**

A bad analogy: Building our pocket   207

Setting up our acoustic pocketing Edit window   208

Pocketing without drums: How to deal with it   208

Dealing with raked chords and trusting musicians   211

Understanding the acoustic waveform and how to
    pocket it   213

Pocketing has left our track early: Pitch 'n Time
    to the rescue again   220

Acoustic pocketing summary   230

Chapter 7   **Electricity in the air**   **233**

Working with electric guitars   233

One at a time: Dealing with three different types
    of electric guitar   233

Changing technique: The evolution away from
    transient pocketing   234

Cleaning up the intro on guitar 1   234

Dealing with dramatically mismatched volume levels   236

Arranging your tracks, pocketing the electric   237

The special guitar: Filling in gaps on a
pocketed electric   238

Our friend masking, and his impact on
electric guitars   242

Spotting the electrics   242

No more peaking: Letting your creativity flourish
when editing electric guitars   244

Identifying and fixing problems using slow
playback and nudging   247

Electric guitar 2: Working with a percussive part   249

Looking for pocketing guides: Parallels in
other guitar tracks   250

A standard electric pocket   253

A lack of time stretching and why   255

The third electric: The ambience track   255

Visualizing the track: Finding places to pocket   256

Spotting misadjusted guitar chords   257

Pocketing against another electric   259

Electric guitar summary   259

Chapter 8   **Autotuning: The not-so-dirty little secret
behind a great vocal track**   **261**

A brief discussion of tuning ethics   261

Pocketing and tuning as a mix, rather than 'fix', issue   262

Starting with our comped track   262

First things first: Getting to know the vocal   263

Setting up the Edit window for vocal tuning   264

Where to start: Selecting your audio to be tuned   268

Graphical versus Auto correction   268

Clearing out the old: Loading in the new pitch
information   270

Tuner settings: Getting the right Retune and
Tracking speeds   271

Viewing your pitches and setting your scale   272

Using tools in the Grid window to correct the pitch 275

Using the Option key to lock in your pitch 276

Undo doesn't work? What do I do to Undo? 280

Falloff notes and how to fix them 281

Listening back: Checking your work 282

Chopping up long notes 284

In defense of AutoTune: How other people
do it wrong 287

Printing the track 288

Tuning special cases and how to fix them 289

The most important part 290

Tuning backing vocals 294

The Chromatic scale and special tuner settings 295

Setting up your backing vocal tracks for tuning 296

Wrapping it all up: A brief farewell 301

Index 303

Brady Barnett is co-owner of Of Sound Mind Productions in Nashville, TN. After working as an engineer and editor in Los Angeles for a number of years, Brady moved to Nashville, where he has spent the past 6 years working as an independent producer, recording engineer, composer, arranger, and one of the premier Pro Tools editors in the music industry.

Brady works with many of the top producers in Nashville (Dann Huff, Keith Stegall, Buddy Cannon/Norro Wilson, Frank Rogers, Brown Bannister, and Pete Kipley), and his impressive client list includes such recording artists as 'N Sync, Faith Hill, Alan Jackson, LeAnn Rimes, George Jones, Reba, Lonestar, Keith Urban, SHeDAISY, and Steven Curtis Chapman, to name just a few. Accomplishments include work on the Grammy award-winning album *Cold Hard Truth* (George Jones), 11-times platinum album *No Strings Attached* ('N Sync), and numerous gold, platinum and multi platinum records.

Nathan 'Adan' Adam, Associate Chair of the Department of Recording Industry at Middle Tennessee State University, has a background in both educational and recording technology. Nathan has worked as a freelance recording engineer and studio technology consultant for a number of recording artists and studios, including Caravell Recording Studios, Hall of Fame Studios, Digital Planet, and even the Hit Factory. He has also been featured as an audio recording expert on MSNBC's *Tonight with Deborah Norville*, and published in *Pro Sound News*.

While working as a Sales Engineer/Consultant for Sweetwater Sound, he equipped Pro Tools studios for recording artists such as Kid Rock, the Insane Clown Posse, Semisonic, Dennis Jernigan, and producers for artists including Reliant K, Sugar Ray, and Creed. He is Pro Tools Level 3 certified and has won several awards for his developments in educational technology for Recording Industry students.

Pocketing, editing, and tuning: The what, whys, and hows of major-label Pro Tools editing

Welcome: How this project came to be

Hi and welcome to Multi Platinum Pro Tools. I don't know exactly who you are, or why you've picked this book up, but I can guarantee one thing: You're about to join us for a very unique experience. We have designed this text and DVD to take you through a learning experience unlike any you have ever experienced on the subject of the industry-standard recording software, Digidesign Pro Tools.

Like many of you reading this book, I (Nathan) am a product of the digital audio workstation (DAW) generation. I started as a musician, moved on to a self-contained drum machine and MIDI sequencer, and quickly found myself exploring the mystical world of computer-based audio. As a self-proclaimed computer geek, it seemed only natural to want to learn more about how and why the world of music recording (at the time) appeared to be moving from the hallowed halls of the million dollar consoles and $50 000 tape machines, to the $2500 computer and audio card. Also like many of you, I happened to grow up in an area that was not exactly a Mecca of recording studios, musicians, and free recording advice. As a result, when I wanted to learn about even base-level topics like what a compressor does, or the parameters on an EQ, my primary sources of information initially came from reading books, or the good old-fashioned Internet. Over the last decade or so, I've found myself exploring or employed on all sides of the recording equation from freelance engineering and production, to recording studio sales, support and, ultimately, education.

I have now been teaching audio production and engineering for nearly half a decade. From the smallest 2-year recording program in the state of Kansas (and believe me, that's small), to the largest recording program in the world (1400+ students in the major), I have worked with hundreds of students, engineers, producers, and other educators to figure out how to better help people learn about the constantly changing world of music production.

When I first met Brady Barnett, students would come out of his class raving about his speed, skill, and ruthless efficiency when it came to using Pro Tools. The ones who got into his classes learned at the end of a knowledge firehose. The ones who didn't get in waited in line. After getting to work with Brady and seeing the experience he brings to the learning table, I just knew that there were thousands of semi-experienced Pro Tools users out there who are feeling like they've taught themselves all they can in their editing journey, and would jump at the chance to watch over the shoulder of a real professional. Multi Platinum Pro Tools is here to provide that opportunity.

The times they are a changin'

Let's face it, the music industry is going through a bit of a change recently. From Billy Idol to *American Idol*, from showtunes to iTunes, the advent of new technologies and paradigms has revolutionized everything from A&R scouting and artist development, to the distribution and promotion process. All the way back to the recording studio, at the major-label level the recording process has moved well beyond the simple matter of hitting Record and pushing faders. Whether you have the singing talent of Jessica Simpson, or the lack of it (like, oh – forget it), there are hidden tools and processes working behind the scenes to make every major performance perfect, and every note exact. It doesn't matter whether you're producing punk, pop, rock, or even country, virtually every major record now goes through a variety of stages well beyond the old days of recording, overdubbing, and mixing.

Oh, don't get me wrong. Most major recordings are still achieved with the traditional tools like a great song, great musicians, and a killer hook. The fact is, you can learn about most of those techniques from any number of outstanding textbooks, videotapes, or even in your local music technology program (check out Focal Press's great line of audio production texts). There are hundreds upon hundreds of experienced engineers, musicians, and authors that are all too happy to teach you about every special technique they have for getting a great kick drum sound or developing the perfect guitar tone. Others work to train you on how to mix a hit record, or even basic approaches to mastering in your own home studio. Why are these types of lessons so easy to come by? Many of these recording, mixing, and mastering tools have remained virtually unchanged for the last 20 years, and the secrets are widely known.

Now, however, fresh technology has changed the playing field.

Ever notice that virtually all current records sound musically flawless, but still somehow human? Each performance is perfect, and yet it hasn't lost its inherent vibe? What can explain this strange phenomenon? Have musicians suddenly improved? Did evolution finally catch up to our vocal cords?

This is where Multi Platinum Pro Tools comes in – here's the secret.

Every major-label recording goes through a variety of finishing processes by highly experienced professionals with a single-minded goal: To make a (hopefully) already great performance into the best performance imaginable. It's that simple. An entirely new layer has been added to the recording process that takes everything and everyone from the great recordings by modern artists, to the lesser skills of their wannabe siblings, and brings them up to musical snuff. What is not so simple is just how much time, energy, and work it takes to get any and every record from point A to point B. Over the course of this text and the accompanying DVD, we're going to expose you to these recently developed (and continually evolving) stages of the modern recording flow that you may or may not have ever even heard of:

- Pocketing
- Time stretching
- Comping
- Autotuning, etc.

This is what Multi Platinum Pro Tools is all about.

A bit of history: The progression of performance creation using studio technology

Ever since the days of recording direct to vinyl, wire, or acetate, musicians and engineers have sought out ways to create increasingly superior recordings that stretched the boundaries of modern technology. From the creation of analog multitrack tape machines (that allowed the miracle of overdubs, and recording different instruments at different times!) to the development of synchronization techniques, engineers have developed method upon invention, idea upon discovery in the quest for making better recordings. Pioneers from Les Paul and Rupert Neve, to Sir George Martin and the Beatles, all developed methods of creating performances that were bigger than just the people playing the instruments.

When we think of great performances, ranging from Queen's *Bohemian Rhapsody* to the latest tearjerker from Shania Twain, we can see that, in many ways, the production process has become as much a part of the performance as the performers themselves. And while this isn't necessarily the whole case for all great performances at all times, to be well versed in the creation of modern music one must be well versed in modern production techniques.

And so appears the nonlinear, random-access, shiny box digital audio workstation (DAW). Some say it's the great musical equalizer, making quality recordings available to the masses. Some call it the death of great music. One thing we can all agree on is that every tool from the lowliest Mbox to the most tricked out HD3 system has enabled anyone who wants to get into producing music to be able to do so. The one thing these tools can't do is teach you how to use them in a musical, creative, and efficient manner. They also can't teach you the best way to work with the musicians, artists and performers to achieve the best performance you can get. That's what we're here for. To teach you how to use several modern tools, from Pro Tools' editing functions, to Serato Pitch n Time, to Antares AutoTune, our single-minded goal is to show you how to take great performances and make them their best.

The 80/20 rule, and how it applies to you

If you've been around long enough, you will have heard of the 80/20 rule. It's a general, common-sense truism that can be applied to a myriad of circumstances and situations. It often goes that '20% of a group will do 80% of the work', or 'You can get 80% of the way to a great sounding album and only spend 20% of the time and money as the big labels'. While cheaper and better quality technology has indeed enabled us to get closer and closer to a major-label sound, there is no substitute for spending that extra 80% of time to get your record up to 100%.

So, this book is really not for the 80% of people who want to just hit Record, slop out an album, and call it a day. It's for you, the 20% that want to take your record to the top of its game and beyond. You are the ones who will help make the great records of tomorrow. Good luck, and happy editing.

Why do we need another Pro Tools book?

There are literally dozens of books, CD-ROMs, and even a DVD or two available that will purport to teach you how to go from a Pro Tools beginner to a Pro Tools master in just a few hundred pages and a few sample audio clips. So what makes MPPT stand out from the pack?

Let's examine the alternatives. In my experience, and the experiences I hear about from the hundreds of recording industry students I work with every year, each of these other resources tends to fall into one of three categories:

1 Mildly useful rehashings of the Pro Tools manual. They may be formatted in a slightly less dry manner, but are ultimately not very helpful.
2 Useful introductions to only the newest of DAW beginners.
3 Helpful tutorials that show you a few good tips and tricks, but bury it under a mountain of redundant information about 'software installation on the Mac and PC', or the meaning of every preference.

The most common complaint I hear is that these resources will contain a mine full of 'I already know that' coal, for every golden nugget of 'Wow, so that's how they do that' advice. If you have successfully installed and used Pro Tools for any amount of time, you are generally past 95% of the counsel located in these resources. Here's what sets Multi Platinum Pro Tools apart from all the rest.

Where's the information coming from?

Rather than being 'just another Pro Tools guide' developed by professional authors and Digi-certified clinicians, every single step in MPPT is straight from the mind and experience of one of the top Pro Tools editors on the planet. Brady Barnett has edited literally dozens of gold and platinum records in a broad range of styles, from Rock and Pop, to Country and Contemporary Christian. From recordings by artists like 'N Sync and Keith Urban, to Reba McIntyre and Faith Hill, Brady is the penultimate end user, with years of experience to show you how he works with Pro Tools. As a result, we won't be showing you every single function and feature of every tool in the PT software. You can read the manual for that. MPPT is here to show you how to work like one of the best in the business, and develop your own techniques from there.

How's the information presented?

Do you fall more in the 'learn by reading', 'learn by watching' or 'learn by doing' crowd? With a plethora of methods available for learning Pro Tools, you need to know which one works best for you! So many Pro Tools educational resources force the 'watchers' to learn by reading, or the 'readers' to learn by watching. And virtually every other resource completely leaves the hands-on 'doers' in the cold! We believe that there's no single learning style that fits every student, and so we have created MPPT to reach out to every type of student who wants to develop these skills in their own time, at their own pace. By combining an interactive DVD, a real Nashville recording session, and a jam-packed, fast-paced text that takes you inside one of the top editing minds in the professional editing world, you will get to watch, listen, and learn as you edit side by side with multi platinum editor/producer Brady Barnett in a real Nashville Pro Tools session recorded by A-list session players!

Definitions: A familiar-feeling guide to some unfamiliar terms

Pocketing

Pocketing is to audio what quantizing is to MIDI. When you play it in it's not perfect. It's human, which is good. When you quantize a MIDI performance, it take a lot more work to make it still sound natural and musical, as opposed to mechanical and rigid. Pocketing is doing that with audio.

Say your drummer's timing is not great, and there's a fair amount of drifting around the beat. Some of that is cool, as you have the natural give and take of a live performance. But too much flex and it's just plain wrong. In other words, there are times to make it right and times to leave it alone. In pocketing, our objective is to go in, put things under a microscope, and fix what's wrong – without taking away what's musical.

A lot of people try to pocket to a click, and make sure that they've edited everything exactly to the beat. In this editor's opinion, that can really kill the musicality of the performance. My policy is to go in, listen to it, and if it bothers you, fix it. If it doesn't bother you, leave it alone. The old cliche holds as true in editing as it does in life.

The trick in pocketing is when you're tweaking things, each edit is only going to make a tiny, tiny difference by itself. As a result, your naked ear may not hear a big difference right as you start the process, especially if you're starting out with tracks done by great players to begin with. But, when you open it up and look in really close, you start to see a lot of things that are just a little wrong, a bit too 'out'.

Take the song *Home* on the accompanying DVD, for example. While it is a great song, with great players and a great hook, it just sounds a bit too 'anxious' to me. Almost like you're leaning into it. And yet, it's supposed to have a relaxed feel. When you get in close, you can see and hear that a good majority of the instruments are right on top of the beat. When we start pocketing, you'll see that 'pocket' becomes synonymous with 'feel' or 'vibe'. As a result, we're taking our song and bringing it back into the relaxed vibe that we wanted for it.

Vocalists, for example, will sing right on top of the beat 99 times out of 100. And anything too ahead will always make a track feel worse than something that's just a bit behind or late. Too much will lead to loose and sloppy, but ahead virtually always feels wrong.

The process of pocketing a typical track might go like this. First, listen to your track. Really evaluate the feel of the song and the message it's trying to convey. Next, look at the performance. Is there anything about the performance that's taking away from the feel of the song? Now, look at whatever your driving instrument is. In most modern songs, this is going to be the drums, and it's going to lead the charge. Once you've pocketed the drums, all other instruments will be pocketed against them, generally slightly behind the drum beat (late).

While the ultimate pocketing experience is one where you start with great tracks and make them even better, there may also be times when you have to use these skills just to bring a poor performance up to snuff. Whether that's fixing a drummer who just can't play to a click, sliding an anxious bass player back into the groove, or mopping up after a sloppy lead guitar, your job as an advanced editor is ultimately about creating memorable performances. And if you keep the music, performance, and the message ever at the forefront of your mind, you will always be able to find work. Because hey, after all, we're all human.

Beat detecting

Beat Detective is a software program inside of Pro Tools that was introduced by Digidesign around Pro Tools version 5 (TDM). When fed a percussive piece of audio with large, clearly defined transients, Beat Detective is able to analyze, chop up, and retime the performance so that even a sloppy drum take can be tightened up into a groovy lick. When used properly, Beat Detective can really help speed up the process of pocketing certain types of tracks. On other tracks it can take up more time than it saves because it can misread a transient and rearrange all the audio regions based on its misunderstanding. As you can imagine, finding and fixing the issue can take some time.

Comping

Comping is the process of recording multiple takes of a particular instrument, most commonly the vocals, and ultimately editing a single master performance take from the pieces of the other takes. This has been a common practice for the last several decades of multitrack recording, and has been done since the days of razor blades on tape, to the most modern copy and paste from one track to the next. While we won't spend a lot of time on comping in the text, we'll do a quick vocal comp on the DVD movie. In addition, there are lots of free tips, tricks, and videos at www.multiplatinumprotools.com.

Time stretching

Also commonly referred to as time compression/expansion (TCE), time stretching is the process of using either third-party software or an included Digidesign tool to change the length of a recorded audio region by either slowing it down or speeding it up (hence the term time compression/expansion). In addition to simply changing the speed, inherent in the modern idea of time stretching is that the pitch of the TCE'd audio does not change, despite its new speed. This is used for everything from speeding up the feel of a recorded track that seems too slow, to stretching out individual audio files to fill in the gap left by an edit. We will be doing a lot of time stretching.

Tuning

Tuning is a common name given to any audio editing activity involving the correcting or altering of pitch in a recorded instrument. It derives its name from the first breakthrough piece of software that really created the market for tuning, Antares AutoTune. Not many years ago, there were no tools to help an engineer out if their vocalist could not sing on pitch. More commonly, great recorded performances were often composed of dozens of takes and hundreds of punch-ins in the search for the perfect vocal take that had the right blend of feel, timing, and, of course, pitch.

In a very short time, Antares AutoTune went from being a little-known, kludgy piece of software that was able to work pitch correction miracles, to being a widely known, kludgy piece of software that can work pitch correction miracles. It has become so ubiquitous in the professional audio editing world that the field it spawned has adopted its name, even though many competing pieces of software are now making their way to market. Just like Xerox, Refrigerator, and Kleenex before it, any time someone uses a pitch correction tool (from the oft-raved Celemony Melodyne, to new pitch tools being bundled in every other major DAW, including the free Garageband!), it is commonly known as autotuning.

That said, while so many other DAW manufacturers are starting to bundle pitch correction programs for free with their software, in order to work through the last few chapters of this text and DVD, you're going to need a copy of AutoTune installed and working on your computer. If you don't have the $300–500 needed to purchase a copy of this software for working through the included sessions, you can always download a free 10-day demo from www.antarestech.com/download.

Since every major tuner in the US market primarily uses AutoTune as their main pitch-correcting instrument, purchasing it would generally be considered a good investment and a valuable skill in today's audio marketplace. And no, I don't work for Antares.

One other thing about AutoTune before we begin. The first time a lot of people became aware of the use/misuse of AutoTune was with Cher's wobbling chorus vocal – 'Do you be-li-ee-ee-ve in life after lu---uv!' This was probably the most pronounced misapplication of AutoTune on the radio at the time, and it was in fact so obvious that it was treated as a sort of pop vocal effect, rather than as a corrective measure. This led to about a million other songs coming out with the same wobbling vocal sound before everyone got sick of what became known as 'the Cher effect'. It still shows up in radio cuts on occasion, though usually in a more subtle manner.

If you want to use AutoTune in this fashion, go ahead, but don't believe for a minute than anyone will think it sounds like a natural voice. Maybe the tin-eared legions that get sucked into *American Idol* can't tell the effect from a natural voice, but you should be able too. (OK, I watch it too – love me that Paula!)

Our objective when tuning, just like our objective when pocketing, is not to alter or dramatically affect the performance. The ideal scenario is when we're working with a great track that has a great feel, and we just get to go through and nudge every killer phrase into perfect pitch. If we have done our job right, at the end of the track, no one will ever know we were there.

Using the accompanying DVD-ROM

The accompanying DVD-ROM has two primary components:

1 A 1.5-hour screen capture movie designed to visually and audibly walk you step by step through the processes of pocketing, comping, and tuning all of the major instrumental and vocal sections on the DVD.
2 Pro Tools sessions with all of the unedited tracks you need to start developing and practicing your own pocketing, comping, and tuning skills.

To use the DVD, simply drag the included HomeDVD_Tracking Session folder, and the QuickTime movie file, from the DVD to your desktop or media drive. Once they have both finished copying, you can eject the DVD from the computer so that Pro Tools looks for the song audio files on your hard disk, and doesn't accidentally look for the audio files on the DVD-ROM. Pro Tools won't allow you to run the session from the DVD anyway, so it's best to just take it out of the computer once you've copied its contents to your hard drive.

You'll need QuickTime 7 to be able to play the video file back, so if you have an older version you'll need to download the free upgrade from www.apple.com/quicktime. It's available for both Mac and PCs, and will let you get your editing groove on by following along over Brady's Multi Platinum shoulder.

Depending on your learning style, note that the information presented in the text and DVD differ slightly in approach, presentation and content. If you're the "follow along while watching and listening" type, and just want to practice editing tracks alongside the QT movie, the movie file will take you step by step through the included session and cover all of the major concepts and practices in the book. On the other hand, a few times in the text we wanted to go a little beyond the topic at hand to introduce you to a few other Pro Tools concepts we think will be useful to you, but that you won't necessarily follow along with in this particular session. These sections will be clearly notated so you don't try to look for tracks or files you don't have, but they'll expose you to some of the practices you'll encounter on your editing journey.

Finally, if you want to compare your edited tracks against Brady's pocketed, comped, and tuned tracks, point your browser over to www.multiplatinumprotools.com to download the final edited Pro Tools session. Then, you can import the edited tracks into your own session, and see just how well your skills measure up!

If you run into any technical problems along the way, again check out the Frequently Asked Questions links over at www.multiplatinumprotools.com for advice, forums, and technical contact links. We'll be sure to set you straight.

So, assuming that you now have either the demo or a full install of AutoTune on your system, the included Home_DVD Tracking folder pulled onto a valid Pro Tools drive, and the QuickTime movie running and ready to go, we're ready to get to work!

Adjusting your seat-belt: Setting up Pro Tools preferences for a streamlined edit session

How to approach a session: Doing things the platinum record way

Now that we've addressed the what's & why's of the MPPT editing process, it's time to get our hands dirty, open up the session, and learn to do things the platinum record way. In my classes, I'm known as the king of bad analogies. That said, I believe that the clearest way to describe this process is to reach out for an unusually (for me at least) good parallel.

Whenever you sit down to work at a new PT system, you need to set it up for maximum recording/editing/mixing efficiency. I like to think of it along the lines of prepping for a long cross-country drive. Just think about it for a second and I believe you'll see where I'm going with this. If you're like my wife, you prep for a trip by checking your oil, tire pressure, adjusting your mirrors, and making sure you've got a $50 tank full of Saudi's finest. You buckle your seatbelts, make sure the CD collection (or iPod) is close at hand, and by doing all of these things a good driver will generally increase their odds of an enjoyable, problem-free road trip in the process.

Then there's the rest of us. If you're like my friend Charlie, preparations for a road trip involve leaping into the car, slamming it into gear, and roaring onto the highway at 15 mph over the speed limit – without ever buckling up. Cars in front are treated as hazards to be passed, while cars behind had best stay there – or else. Lights on the dashboard and strange smells emanating from the engine are considered trivial until the transmission physically falls out onto the ground.

Now, who do you suppose is more likely to encounter frustration and problems along their journey? I'll let you sit on that (no pun intended) for a minute.

Configuring your session, choosing your settings, keeping your sanity

To edit like a professional, you have to approach your Pro Tools editing sessions with a broad body of knowledge. You need knowledge of music, knowledge of the system, and obviously an extensive knowledge of the tools at your disposal. Knowing how to set up and configure those tools to behave in a manner most efficient to you is of critical importance as well.

One great thing about Pro Tools is that with every new version it becomes just a bit more agile, flexible, and versatile, and every engineer is able to configure it to behave in a way that seems most logical to them. While some may argue that Pro Tools is the least configurable of the modern DAWs on the market, the fact that it is unchallenged in its industry-leading role clearly shows that Digidesign is doing something right.

The flip side is that it's a double-edged sword for the relatively new engineer who has diligently learned everything their manual can dryly, slowly, painfully teach them about their shiny new Digi002, Mbox, or even home HD system (what – you don't have an HD system in your bedroom???).

For example, thousands of up and coming engineers have been teaching themselves to record, edit, and mix in Pro Tools with whatever options were checked by default the first time they launched their system. This is all well and good when you are independently wealthy, and able to force all your work to come to you. As for the rest of the world, what happens when they get called out to edit in another studio, in another city, on another Pro Tools system that that has been set up quite differently? What happens when the whole system looks, feels, and behaves in an alien fashion relative to what they're accustomed to? Chaos and mass destruction – that's what. Well, it may not be that bad, but it's probably at least as bad as getting into your car and finding someone has changed all the radio presets – it's a sad, sad day.

Solo buttons that used to latch no longer latch. The record arming buttons only allow you to record on one track at a time. The playback head starts and stops as though possessed by the evil ghost of an analog tape machine, and Pro Tools mockingly ignores audio selections you've made with a cold, bitter shoulder of indifference. Worst of all, when the best vocal comp is finally finished, and the band comes in to hear their song that will be heralded as changing the face of modern music, the spinning pinwheel of death brings the operating system to a thunderous crash, and all the Autosaves you thought were being made, unfortunately, were not.

Each of these horrific scenarios is a quick and easy way to bring a recording or editing session to a vibe-killing halt. In addition, for a newer engineer who has not learned to keep a cool head when things start going wrong, these scenarios can start to pile up on one another until the frustration reaches boiling point. The simple fact of the matter is that every one of these scenarios can be avoided by simply knowing what individual options and preferences need to be selected to make Pro Tools operate in a manner most efficient for an editing session.

A caveat – as this text is primarily about multi platinum editing techniques, know that you might choose to set a few of these preferences differently during, say, the recording, mixing, or mastering phases. These are just the preferences that work best for one multi platinum editor, and we'll always tell you why we think they're the best, and why anyone doing it another way is wrong (just kidding!).

In addition, we're not going to cover every single preference located under every tab. For example, it really doesn't matter what color coding scheme you want for your tracks and regions (unless you're Martha Stewart, in which case might we suggest a nice black and white striped affair?).

Pro Tools preferences and how they affect your session flow

After copying the session from the DVD to an empty spot on your hard drive, go ahead and launch it and let's work through all of the preferences, options, and views that will make you a lightning fast editor.

Figure 2.1   Start with the Preferences located under the Setup menu.

If you've been using Pro Tools for a while (and we're operating under the assumption that you have), you've probably nosed around in this box before, and maybe even know what a few of these options do. I've found, though, that most people have never given a moment's thought to any preference that doesn't seem to directly affect what they do on a daily basis. I liken this to a person that doesn't bother to learn about changing their car's oil until their engine comes to a grinding halt. See the previous section for more bad analogies.

Once your Pro Tools Preferences box comes up, we're going to start with the tab that has the most dramatic effect on your editing session. Can you guess which tab that's going to be?

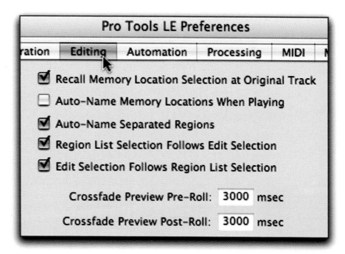

Figure 2.2 The Editing Preferences tab affects your – drumroll – editing preferences!

There are several preferences in here that you need to confirm before you get the session going.

The first one I always turn off is the **Auto-Name Memory Locations When Playing**. Because I'm prone to getting off the subject, let's start by noting that Digidesign uses the longer term 'Memory Locations' rather than the more common 'Markers' because Markers are only a subset of Memory Locations. A Memory Location in Pro Tools can refer to a selected region or set of regions, as well as Zoom levels, or location times. Neato. Regardless, I prefer to leave this preference off. As a track is playing back during the edit, or even when the band is first playing through the track, I like to hit the Enter key (on your keyboard's Numeric keypad) at all the major song sections, and type in a quick label for verse, chorus, bridge, etc. Hitting Enter again will place the marker in the song at the location you first hit it.

This is also useful for beginners who might benefit from listening through the track and dropping markers (ahem, Memory Locations) into parts of the song that obviously need tuning, pocketing, sound replaced, etc. When doing an actual record, we will physically be going through every note of every track in the song anyway, so this isn't really necessary. However, if you would benefit from leaving yourself these sorts of visible reminders for the really obvious parts that need tweaking, be my guest.

The other side of the coin would be that with this preference turned on, Pro Tools will give the marker the unhelpful name 'Memory Location 23, 24, 25, etc.'. The way I see it, more organization is almost always better than less, and why would you want to place a marker if you didn't want to know what it's marking? This is not a pirate's treasure hunt, and a good marker is worth its weight in gold for finding a specific location, not to mention all the time it saves you while editing.

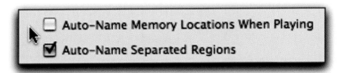

Figure 2.3　Two important Editing Preferences.

The next one to check on is the **Auto-Name Separated Regions** option. This seems obvious to me, but I have heard some people try to rationalize turning this preference off. The way I see it, over the course of editing everything from multiple tracks of drums, guitars, and layers of background vocals, if we had to stop and name every single region we're going to create, it would be literally thousands of extra steps, and hours of wasted time. Beyond this, if you have lots of drum tracks or 48 layers of 'N Sync style backing vocals you could get upwards of tens of thousands of regions fast, quick, and in a hurry. Now imagine trying to keep track of them all (now was that cymbal hit Crash_1928374 or Crash_1928375?).

Below the list of editing checkboxes you'll find the options to set your default **Fade In**, **Fade Out**, and **Crossfade** shapes.

Figure 2.4　Setting your default fade shapes.

These are the fades that occur whether you create them with the Smart Tool or with the Create Fades command – Cmd-F (Mac)/Ctrl-F (PC). Initially, these fades are set to a simple linear Fade In, Fade Out, and Crossfade. Clicking on one will allow you to ramp your incoming, outgoing, and crossed regions with fades of Equal Power, Equal Gain, or the ever-sexy S-Shaped fade.

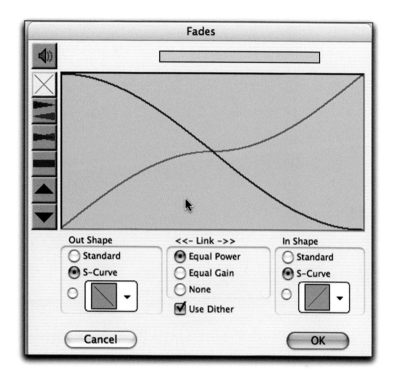

Figure 2.5 A sexy double S curve – that unfortunately you'll rarely use.

'Which one should I use?', you ask with a puzzled stare. That really all depends on the style of music (i.e. classical, rock, country, etc.), whether it is a two-track radio edit or just one layer in a multitrack project. I suppose one other major factor is how picky you are.

Once upon a time I was a meticulous nut when it came to edits and fades. I would solo every track, from toms to snares to tambourines, listening through for the slightest hint of an audible edit. Ultimately, as I worked with more and higher quality engineers, tracks, and producers, I realized that people are (most of the time) way pickier than they need to be with fades. While I realize that certain styles of music (like classical) may require special Equal Power fades to properly make stereo orchestral edits fit in a mix, in a pop/country/rock music track it just tends to be wasted time.

As a result, when it comes to the default Fade In and Out shapes, I find that your standard Equal Gain linear Fade In and Out shapes work really well.

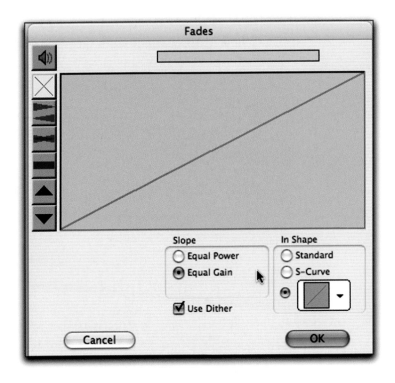

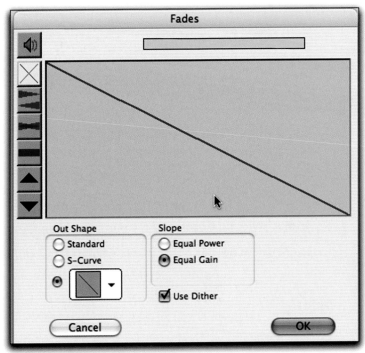

Figure 2.6 Standard Fade In and Out shapes.

Though I may stick with the linear fades most of the time for basic editing, I still spend a lot of time listening in solo. If I hear a lot of bumps in a track I will either go back and fix them, or possibly try something with a more logarithmic shape.

The only type of fade I usually find myself being picky about is the default Crossfade. I like to set it up as follows.

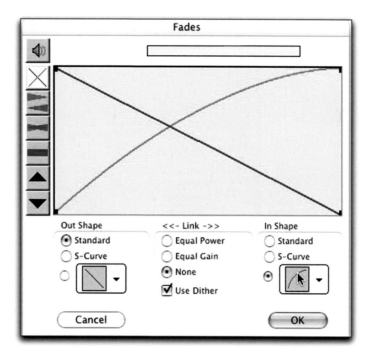

Figure 2.7   My default Crossfade – because it just works.

Set your **Link** option to None, your **Out Shape** to a Standard linear, and your **In Shape** to a logarithmic ramp. After years of editing all types of drums, styles of voices, and sounds of instruments, I have found this to be the best overall shape for working in a more contemporary music setting, for almost all musical styles.

Now that I've given you that tasty little piece of experience that took me untold hours to arrive at, let me snatch it right back with another great big caveat. When editing drums, bass, and other instruments, as you will soon see, you often have to employ the habit of micro time-stretching regions of audio to fill in gaps created during the editing process. For these small regions that have been created during the time-stretching process, I've actually found that a straight-ahead double linear fade creates less bumps or audible artifacts, and should be used instead. For more examples, take a look at drum editing in Chapter 4.

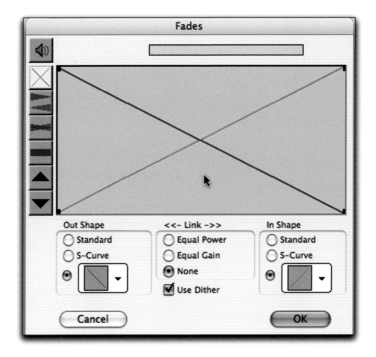

Figure 2.8 The plain vanilla double linear crossfade works best on time-stretched regions meeting non-time-stretched regions.

As a brief aside, for all of these types of fades, and in fact most facets of any DAW, if you see an option to **Use Dither** in just about any situation, it's almost always best to leave it 'on'.

Regarding the length of the fade, for most modern styles you want to avoid a fade that is too long as well. There's no standard for what is too long, as it is going to be a factor of the tempo of the music, and how close the outfading instrument/vocal is to the one coming in. When it comes to vocals, my particular forte, I'm usually editing together a comped track from a variety of takes. If you create an edit with too long a crossfade, you will hear both takes of the word as one fades out and the other fades in. Just listen closely for this and you'll hear how to simply make your fade long enough to hide the edit, while being short enough to avoid detection. Start with a short fade around 4–6 milliseconds, and avoid longer ones approaching half a second or more, unless the track calls for it.

One special situation that cannot be overlooked is if you are working as an editor for a producer/engineer who thinks they can hear the difference on these types of crossfades. If you have the privilege of being hired by someone who trusts you to tune their vocals, or pocket their guitars, find out if they have a preference, or just look at other edits they've already made in the track. When they open it back up after you've finished editing, they want it to look like they edited it, and sound like they expect. In those situations, do whatever the producer calls for. Use their preferred crossfade shape, length, whatever. Even if you can't hear it, you can still take the credit for doing a great edit, just like they would have. You can't lose.

An area we have not yet discussed is the ability to edit without using crossfades at all. To give you a little preview, there are certain times when cutting an instrument that you can totally avoid the need for a crossfade. In fact, often when you are editing two-track material (say you're making a 3-minute radio edit of a 5-minute song), a crossfade will make your edit *more* audible, instead of less. By zooming down to the sample level, and simply lining up the edges of the outgoing and incoming waveforms, you can often avoid any bumps in the audio, and steer clear of crossfades entirely. Intrigued? Check out Chapter 6 on bass pocketing for a quick and heads up demo.

For a more visual explanation, see the accompanying DVD.

Crossfades, drum edits, and hard drives

One of the main places you'll find yourself making truckloads of edits in a very short time is when pocketing drums. Since drums are typically grouped together and any edits are made to all eight, 10, or 12 drum tracks at once, the number of edits needing to be read off the hard drive can become very, very taxing for even the fastest of FireWire or SCSI drives. In addition, since Pro Tools uses file-based (as opposed to real-time) fades, if you're going to be pocketing every note of a drum kit like I tend to do, this creates a 200% increase on the number of files needing to be pulled off the drive simultaneously for it to play back at tempo (the actual audio file, and now also its incoming and outgoing fade).

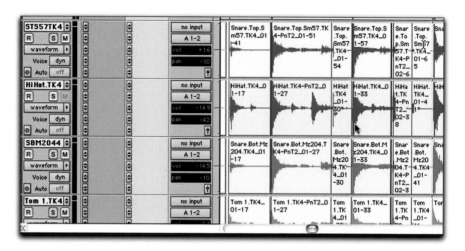

Figure 2.9 Hold off on the crossfades till you've finished editing drums – your system and hard drive will thank you.

Depending on the speed of your hard drive, you may need to take measures to avoid this unnecessary strain on your system (which likely will grind to a screeching halt unless you've got your own superfast terabyte SCSI drive). If you need to, just put off adding the crossfades until you've finished editing your drum track. While you may have to put up with a few clicks and pops as you're working through it, any modern 7200+ rpm hard drive should be able to handle it, and still deliver you quality performance. The best part is, once you've finished, you can simply select all of your drum tracks (Shift-click the regions, or just drag select over them all), and hit your Create Fade shortcut to add fades to every edge, of every selected region, in the entire track. Talk about a time saver!

For more on this topic, check out Chapter 3 on drum editing.

Conversion Quality sets how hard Pro Tools is going to work on sample rate converting any files you import, assuming they are at a different sample rate than the session, of course.

While it probably really doesn't make that much difference if you're just talking about the small samples we might bring in during the sound replacement process, there's no real reason to set it to anything less than **Best**, and I personally happen to be a fan of anything with the word 'TweakHead' in it – you may feel differently.

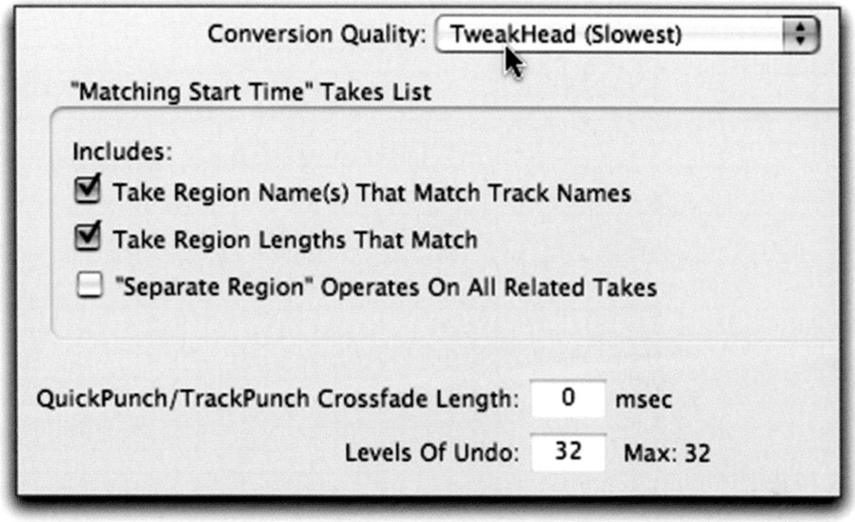

Figure 2.10 Setting your Conversion Quality to TweakHead will yield the best high-frequency response when converting imported audio from other sample rates.

The **Matching Start Time** options are primarily important during the comping phase of production. If you've done multiple takes on a single playlist in Pro Tools, these can make your life a little easier for finding the alternate takes buried under the top one. The **Take Region Name(s) That Match Track Names** would be the most helpful if we were going to comp with the multi-take per track approach, as it will bring up a list of all takes with the same root name as your top region, hence allowing you to quickly audition alternative performances in an expedient manner. However, since I don't generally comp takes in this manner, you can leave them wherever you like.

Since I know many of you may be editing on LE or even M-powered systems, let's address an issue where Digidesign is somewhat behind the curve of other DAW manufacturers. The **Levels Of Undo** option maxes out at 32 levels of undo. While this may seem rather paltry compared to other DAWs, 99, 999, or even theoretically unlimited branches and permutations of life-saving undo (as in Digital Performer, among others), the way that Digi handles the undo functions actually utilizes a fair amount of system resources. As a result, if you are on a slower machine, or experiencing problems being able to play back a certain number of tracks, sometimes lowering this number can actually improve the performance of Pro Tools, at the expense of being able to back out of your mistakes. Depending on your level of experience, it may or may not be a reasonable sacrifice to make. At a certain level, you may be able to edit quickly and only ever find a need for 15–20 levels of undo. However, when you are just starting out, as every edit, nudge, and adjustment will consume one of your undos, you had probably best just leave this all the way up.

Now that we've set up our Editing Preferences, let's define some of the ways Pro Tools behaves under the Operation tab.

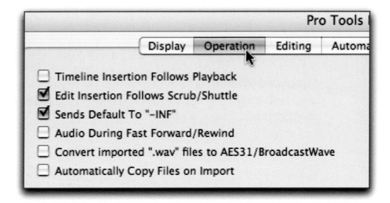

Figure 2.11   The Operation tab can make Pro Tools behave in some very cool or very strange ways.

Many of these preferences can really screw up an editing session if they were set the wrong way, and you didn't know how to correct them. Especially when it comes to pocketing and vocal tuning, one of the worst offenders is the **Timeline Insertion Follows Playback** option. Let's say right off the bat that for editing purposes this option should always be turned off. If checked, the easiest way to describe it is that the Pro Tools transport timeline will behave exactly like a tape machine, with the start and stop points following your playback. Wherever you stop the track, the timeline insertion will be placed there, and when you start it again it will start playing at the same location it stopped at.

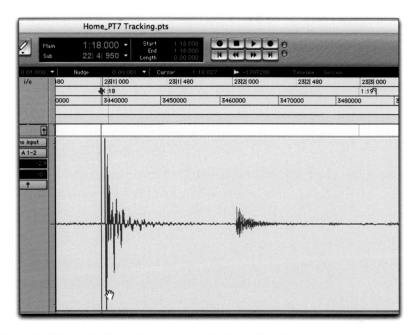

Figure 2.12   With Timeline Insertion Follows Playback turned off, playback starts here. When it's turned on, the playhead returns to where it started from, allowing us to listen closely to this soundbite repeatedly without having to rewind after every listen.

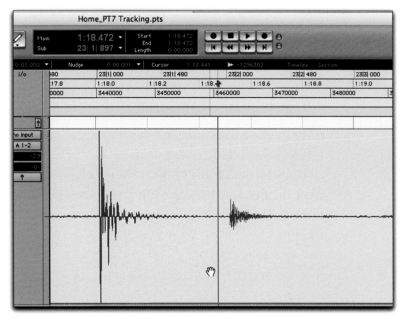

Figure 2.13   When stopped, the timeline cursor follows the playback, stopping and staying later in time, just like a tape machine.

For editing purposes, we don't want Pro Tools to operate like a tape machine. We want it to function like a, you know, nonlinear, random access digital audio workstation, for example. When this function is disabled, Pro Tools allows us to set the timeline insertion at a particular point, starting and returning to that point every time we toggle between Play and Stop. Therefore, for focusing in on a particular edit point that we want to audibly hone in on, it makes the most sense to be able to start and stop at a point of our choosing for repeated listening, without having to rewind after every listen – like we would with a traditional tape machine.

Another preference that will make your editing life easier is the **Edit Insertion Follows Scrub/Shuttle**. For example, when I'm listening to a track and need to get down, dirty, and nitpicky, the Scrub tool allows me to listen in a microscopic fashion to a small sliver of the tracks, and easily discern any out of time/out of tune pieces for termination.

Figure 2.14   Make sure these preferences are set appropriately – you will use them all of the time.

Of course, with the Scrub tool, the tighter you are zoomed, the more focused the tool becomes. Thus, once I have located the offending sound, with this preference engaged my Edit Insertion cursor will stop at the location I scrub to. With this preference unchecked, my Edit cursor will stay at whatever location I last left it at.

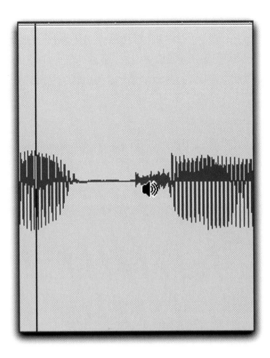

Figure 2.15 With the Edit Insertion Follows Scrub/ Shuttle option turned off, the timeline will jump back to its old position (on the left), even though you scrub to the breath sound you want to delete.

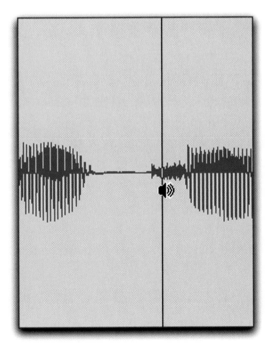

Figure 2.16 With the Edit Insertion Follows Scrub/ Shuttle option turned on, it will stay at the location you scrub to. Cmd~5 (Mac)/ Ctrl~5 (PC) engages Scrub mode.

This is particularly effective when working through a vocal track, where I am often zoomed in very tightly, scrubbing through at the word and syllable level to eliminate unnecessary lip smacks, coughs, acid reflux, and the like – as opposed to all those 'necessary' lip smacks and coughs, that is.

While there are occasional times during the recording (but not editing) process that some engineers might prefer to have the Timeline Insertion Follows Playback preference in either state, I can't think of a good reason to turn off the Edit Insertion Follows Scrub. I don't know what other people may be using Scrub for, but if I'm scrubbing, I'm looking for a place to edit – so I'd naturally like my edit point to follow my scrub, simple as that.

The next column kicks things off with a few preferences that I also tend to leave on.

Figure 2.17 Three options that are best left on, unless you only like to record one track at a time.

The **Latch Record Enable Buttons** shouldn't really even be a choice as far as most music production is concerned. If you turn this off, you'll find your session going very poorly, as every time you record arm a track, it unarms every other record-enabled track. This is like the old days of only being able to record one MIDI track at a time. Bizarre, but you should be aware of it in case it ever gets turned off on you.

Link Mix And Edit Group Enables is also an 'always on' situation for me as an editor. Possibly, if you were using Pro Tools as more of a simple 'tape machine' multitrack and only editing occasionally, I could see someone wanting to de-link the Mix and Edit groups. However, when you're working entirely within Pro Tools on your initial tracking, edits, overdubs, and mix, I generally have tracks like drums, guitars, and BGVs grouped for a reason. After all, one can always disengage the group momentarily if something needs to be adjusted on an individual track.

This is done in the Mix window by click-dragging an individual fader within a group and holding down the Control key.

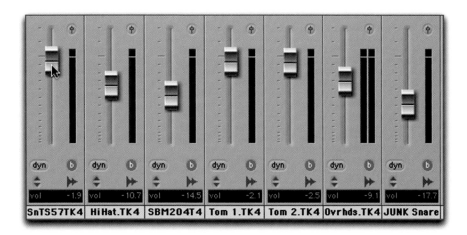

Figure 2.18 Notice the green 'b' indicating all of these drum tracks are grouped together. Moving one fader will move them all simultaneously.

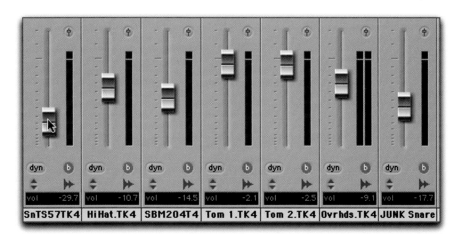

Figure 2.19 Holding down the Ctrl key allows us to adjust the left fader without affecting the level of any of the other faders.

To momentarily suspend the groups for editing more than one fader, or to edit just one track of a group in the Edit window, simply type the letter key associated with the group, with the Keyboard Focus mode turned on. This is set up by default in the Mix window, but in the Edit window you must first select the little A/Z button in the upper right corner of the Edit Groups window for it to work.

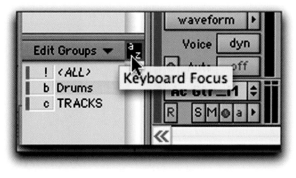

(a)

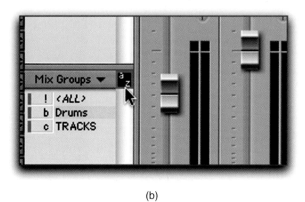

(b)

Figure 2.20 The Keyboard Focus mode must be turned on in the Edit window (a) but is on by default in the Mix window (b). To turn specific groups on or off, simply type the letter located to the left of the group name, i.e. b = Drums, c = TRACKS.

Link Record And Play Faders is also a preference that mostly impacts a session flow during the course of the overdub recording. With this preference, the level the fader is at when the track is record armed is the same level it is at when not record armed. Hence, they are 'linked'.

(a)

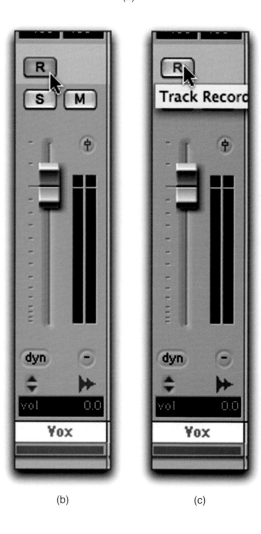

(b)   (c)

Figure 2.21 With the Link Record And Play Faders preference turned on, the faders stay at the same level.

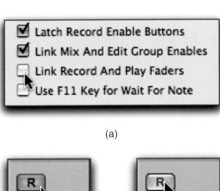

(a)

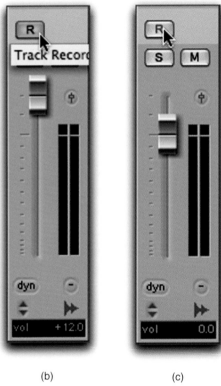

(b)   (c)

Figure 2.22 With the Link Record And Play Faders preference turned off, the faders can be set at different levels for record monitoring and for playback in the mix.

When disengaged, you can set the fader at two different levels. You would use this during, say, a vocal overdub to set the monitoring record fader to a higher level, but when playing the track back (unarmed) the fader would drop back to the already established rough mix level for the vocal.

One preference that should *never* be turned off is the **Enable Session File AutoBackup**.

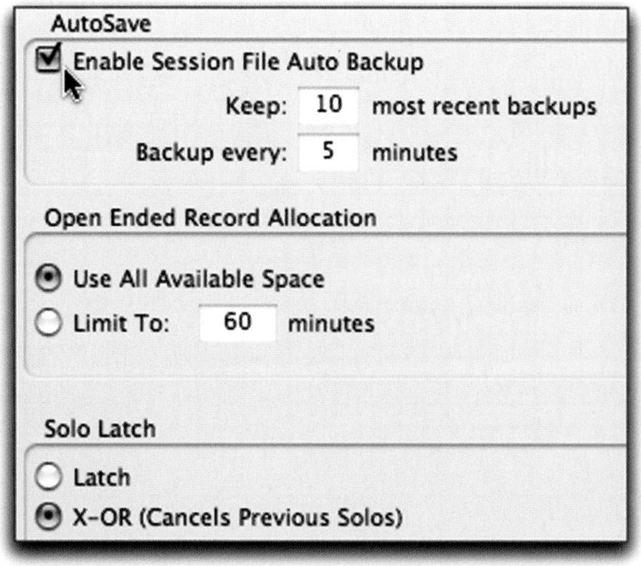

Figure 2.23 AutoSave, Record Allocation, and Solo Latch.

With this turned off, you would have to develop what is widely known in the recording industry as 'the twitch'. The 'twitch' is the reflexive action of hitting Cmd-S (or Ctrl-S on a PC) to cause the software to 'Save' the project – every few seconds. It's a debilitating disease that besets DAW engineers everywhere who have lost hours of work to a lighting strike, because they failed to save it on a regular basis. Set it to backup regularly, say every 5 minutes. You'll never regret it.

Open Ended Record Allocation should virtually always be set to **Use All Available Space**. Back in the days of yore, when hard drives were small and CPUs slow, dropping a bunch of tracks into record and letting the machine run was a sure-fire recipe for pain and heartache. Often, when the computer ran out of drive space, rather than keeping everything it had recorded thus far it simply freaked out and wrote a massive 2-gigabyte file full of white noise, or some other such atrocity. At that time, it was a good idea to limit your freewheeling record time so you wouldn't run out of space and kill everything you'd done up to that point.

With hard drives now sinking below $0.30 a gigabyte, and terabyte disks to be had for less than the cost of an iPod – well, you get the idea.

The last option here is whether to **Solo Latch** or not. This one can actually go either way depending on the situation. With the Latch option turned on, Solo buttons stay – wait for it – latched! With this preference you can go around soloing track after track and it will work as an additive solo mode, adding more and more audible tracks to the originally soloed track.

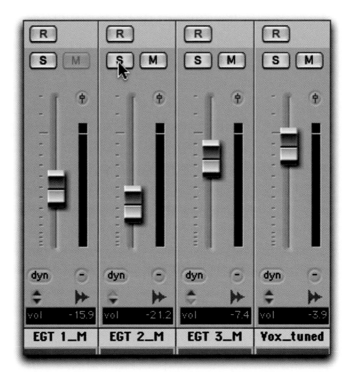

Figure 2.24 Solo Latch just adds tracks to the Solo bus whenever you select Solo on more than one track.

With the Latch option disengaged, only one track (or group, or instrument track, etc.) can be soloed at a time, disabling the previously soloed track every time you click the shiny yellow 'S' on another new one.

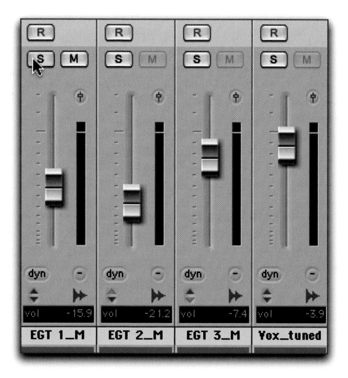

Figure 2.25 The X-Or Solo Latch disengages any other soloed track every time you solo a new one.

During the editing process, disengaging the Latch option allows me to quickly jump back and forth between, say, two instrument takes/tracks, an untuned and a tuned vocal track, or a pocketed and an unpocketed version of a guitar track.

When you first start out learning advanced editing, pocketing, and tuning techniques, it is often useful to duplicate the track you want to work on first, and then regularly listen back to it soloed back and forth against the original track. This helps ensure that you're actually improving upon what you started with in the first place. By unlatching the solos, it allows you to quickly toggle between the two versions of the track to microscopically tune in to your changes and improvements. Once your skills have improved to where you no longer need to regularly do this, I would advise you to leave the Latch Solo option turned on. It's a slightly more standardized way of working.

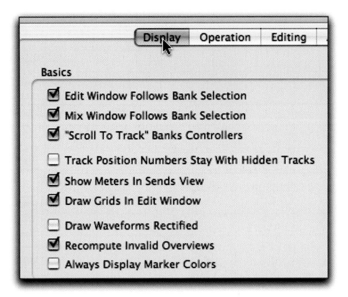

Figure 2.26   Click over to the Display tab to change how things appear.

Moving at a turtle's pace over to the Display tab, you'll find a whole passel (that's southern for a lot) of preferences for how your tracks, metering, grids, plug-ins, and waveforms are displayed in Pro Tools. Ironically, after all that discussion about knowing what bad preference choices can do to your Pro Tools setup, only a few of these preferences will likely have an impact on your editing session.

Some, like the **Edit/Mix Window Follows Bank Selection** and **"Scroll to Track" Banks Controllers**, really only apply to the mixing phase if you're using a control surface like the Command 8, Digi 002, or the Digidesign ICON. Others, like **Default Track Color Coding**, **Peak Hold**, and **Clip Indication** are really just a matter of personal preference.

The first one to check on (pun fully intended) is the **Draw Grids In Edit Window**. While you won't necessarily need this to function well in an edit, having the gridlines showing on the screen helps some people have a better feel for where the notes they're pocketing are in relation to the beat, assuming you tracked to a click, that is.

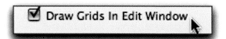

Figure 2.27

The other side of that preference is that even if the session was recorded to the PT click, the whole process of pocketing, done properly, is not about lining everyone up to exact notes or some mechanical grid. That would take all of the humanity out of the performance. So, I tend to only use the Grid and Gridlines for selecting portions of a song for flying around (i.e. copy and pasting) from one section to another, or selecting sections to delete in maybe a radio edit. Ultimately, you will develop your ear and eye to the point that you likely won't need this preference hardly at all. When you're first starting out, though, it can be a real lifesaver.

To look at a few more preferences that I tend to like one way or another, the **Organize Plug-In Menus By** option is one I tend to turn to **Flat List**. In older versions of Pro Tools, all the plug-ins appeared in one giant list every time you clicked on an insert. The categories are useful if you happen to know exactly what category a plug-in falls into (like a Waves C4 under Dynamics, or a JOE MEEK Meequalizer under the EQ section), but for many plug-ins that might not neatly fit a category (Vocoders, Pitch Shifters, Stereo Imaging, etc.), it can be a bit of a pain to hunt through sub-folders to find what you're looking for. This one goes either way though, really. Whichever works best for you.

Figure 2.28   My personal preference for viewing plug-ins.

The last Display preference that really makes a difference when it comes to editing is **Draw Waveforms Rectified**. I always leave this off. The fig-ures below show the difference between a standard waveform and one drawn rectified.

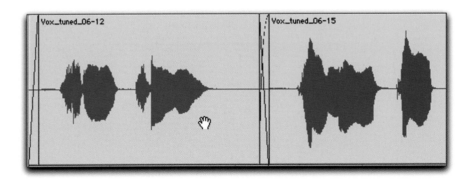

Figure 2.29   A standard waveform – good for editing.

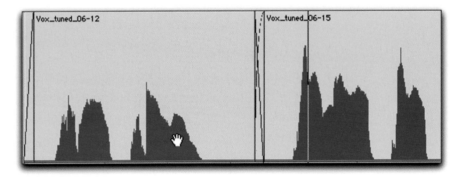

Figure 2.30   The same waveform rectified. This is a non-standard way of viewing waveforms, and isn't the best for pocketing and editing.

Some engineers prefer a rectified view when they're working with lots of audio tracks on the screen at once, as it allows them to see more of the waveform by summing the positive and negative sides together. It really doesn't work best for editing though, so leave it off.

Really, most of the other preferences under the Display tab are relatively harmless. You can turn on your **Tool Tips Display**, or not, and choose to have marker colors or not.

The last preference we need to check is located under the Processing tab, and it's labeled **TC/E**, standing for Time Compression/Expansion.

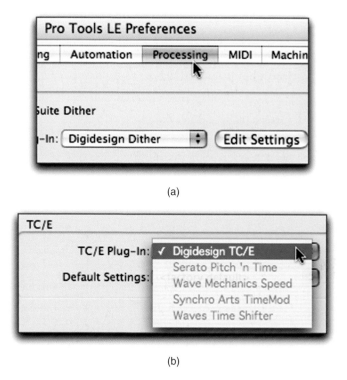

(a)

(b)

Figure 2.31 Find the Time Compression/Expansion plug-in under the Processing tab.

We will discuss later how and when this setting should be changed from its default setting of Digidesign TC/E. For now just know that it's here, and that under the default setting you have the option for a variety of TC/E plug-ins available from third-party manufacturers.

Before we move on, don't forget that, at least in current versions of Pro Tools, preferences are local to the system, and not the session file. While some DAWs handle this differently, with Pro Tools, any changes you make to the preferences on your current system stay there, and do not travel with the session file. Got your home setup just the way you like it? Great. If you move the session to a rig at another studio you'll need to set up all of these preferences from scratch. This is a compelling reason to learn what all of the important preferences do, because if anyone else ever uses your system and changes things around, it can be a world of anguish putting things back to the way they were, especially if you're not sure exactly how you had them previously.

Now that you've got a working knowledge of all the critical preferences for a successful edit session, go ahead and click the big blue Done button at the bottom of the Preferences screen and let's look at how we're going to configure our Pro Tools screen layout.

Figure 2.32   Just because I like graphics.

Mix window, Edit window, or both? Why use two when you only need one?

If you grew up in the stone age of digital audio workstations, you may recall that early versions of Pro Tools actually had two separate applications for mixing and editing. To make a very long story into a single short sentence, these ultimately evolved into the Mix and Edit windows that we know and love so much – except that I don't.

While I can appreciate the old standby model of having a screen that looks like a mixing console to comfort the weary engineers who are just now moving into the digital age (welcome, by the way!), the fact is that you not only don't need it at all for the editing and overdub phase, but I really never utilize it for the recording stage of a record. Here are the secrets.

What do you ever flip over to the Mix window for anyway? Faders? I/O? Buses, inserts, comments? All of these are accessible within the Edit window in just a few short clicks. Now, assuming you're working on a reasonably sized monitor that will still leave you room for editing, let's set up our Edit window to be the fulfillment of our editing dreams.

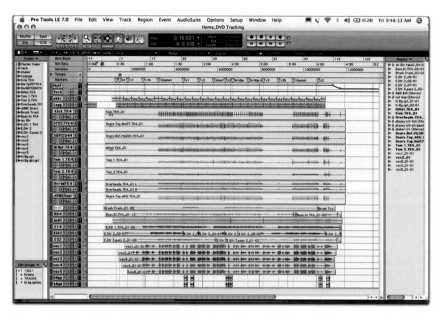

Figure 2.33 The default Edit window. Show/Hide and Groups on the left, Tracks in the middle, and Regions bin on the right.

The quickest way to do this is found under either the **Edit Window View** selector directly underneath the Slip mode button, or under the new View menu in PT 7.

Figure 2.34 Setting your Edit window up with everything it needs.

Under either of these options go ahead and turn on the **Inserts**, **Sends A–E**, and **I/O** options. Depending on if you are using the new instrument tracks or the extra five sends just introduced in version 7, you may also want to turn these on as well, but for editing purposes they'll likely just be in the way.

(a)

(b)

Figure 2.35 Choosing these options will keep all the necessary components under one window.

Now that those options are engaged, I also like to start an edit with all of my tracks set to a medium height. While this will obviously change constantly, it's a good default height to go back to on a regular basis, as it quickly allows you a few more creature comforts, as you'll soon see. Go ahead and hold down the **Option (Mac)/Alt (PC)** key on your keyboard, then click on the ruler between your I/O and Tracks column to make this change.

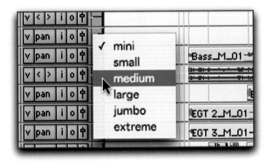

Figure 2.36 Option-clicking anything in Pro Tools affects *all* tracks, including resizing them all to medium height.

Link Timeline and Edit Selection

The last option you need to check is located under the newly revised and eminently better Options menu, called **Link Timeline and Edit Selection**. (Not to get too far off the subject, but to those of you who've been using Pro Tools for a while, I just really enjoy the new menus and their layout in version 7. Once you get used to them, they really make a lot more sense than the old way.)

The Link Timeline and Edit Selection will do just what its name implies. Anything you select for editing on the timeline, from a region to any time selection, is also what will be played when you start playback. Most of the time when we are selecting something during the edit, we will want to listen to it repeatedly. If we want to hear before and after our selection we will just turn Pre/Post-Roll on and off. Without this selected, where Pro Tools starts its playback will not be related to the audio clips we have selected. I virtually always work with this option turned on.

Figure 2.37   Link Timeline and Edit Selection.

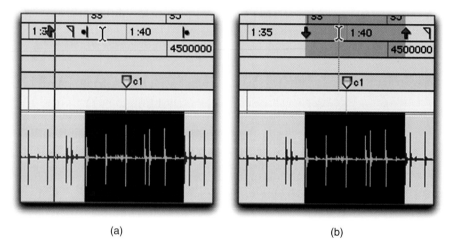

(a)   (b)

Figure 2.38   Link Timeline and Edit turned off (A) leaves the timeline unrelated to the selection. Turned on (B), the selection is what plays when you start playback.

Utilizing the Show/Hide button

The last step is to utilize the Show/Hide button to hide the Regions List, and show both the Tracks List and the Groups on the left of the screen. We lose the Regions bin because its really not used that much during the editing process, and hiding it frees up a good portion of the screen for editing.

I personally like to leave the Track List open because we will regularly be showing and hiding lots of tracks as we work through pocketing the various instruments. For example, when pocketing the bass guitar we will tend to hide everything but the kick, snare, and possibly the loop/click. This will help us both visually and audibly focus on the specific instrument we're trying to work with. The Group List will be visible along with the Tracks window because we will be engaging and disengaging groups regularly as we edit.

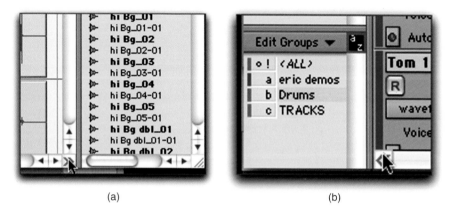

(a)   (b)

Figure 2.39 The Show/Hide button in the lower right corner of your screen frees up screen space by hiding the Regions bin. The one on the lower left opens up to show your Tracks Show/Hide box and Edit Groups.

Checking all of these preferences and options should leave your Edit window layout looking something like this, with the **Show/Hide Track List**, **Group List**, **Track Names/Record/Solo/Mutes**, **Inserts**, **Sends**, **I/O**, and **Tracks** visible on the screen at medium height. Don't worry if the audio tracks on your screen aren't in the order below – we'll be changing all of their order/visibility when we start the edit anyway.

Figure 2.40 The Edit window with all the necessary bells and whistles.

With the preferences set and the screen laid out for an expeditious editing session, let's look at just a few of my personal preferences for methods and tools I employ to get my work done, and get it done fast.

Slip, Spot, Shuffle, Grid: A brief look at the major editing modes and tools

Now, I can already hear a few of you complaining. 'You said this is going to be nothing but advanced Pro Tools! Who needs to talk about the four modes and the editing tools!?!' I know, I know, and you're right. I said this book will assume you've already got an operational understanding of Pro Tools, including the four modes and all the major tools. On the other hand, completely fabricated statistics show that there are going to be a large portion of you who are opening the program for the first time, and just bought this book because the title art looked cooler than any of the other books on the store shelf. Well, thank you for succumbing to our marketing ploy, and I'll try to make the next few paragraphs as quick and painless as possible.

Unlike so many endeavors to explain the modes and tools separately, let's take a look at them together, as they really are forever linked in their usefulness. To put it in very plain English, the four modes affect how the audio regions will behave when you are using your tools on them. Whether you're chopping and looping drums, flying guitar solos from the beginning to the end of the song, or have made a huge mistake and need to put the audio parts back where they were originally recorded, the four editing modes will make your life a lot easier and often pull you out of a jam, and you don't even have to pay them extra.

Slip mode

Describing the four modes as it relates to my own personal workflow, I can start by telling you that you will want to be in **Slip** mode over 90% of the time. To demonstrate this, select the Grabber tool and grab any region in the currently open session.

Figure 2.41 When in doubt, put Pro Tools in Slip mode. Shortcut key F1.

Figure 2.42 The Grabber tool is handy for grabbing and moving regions. Shortcut key Cmd~4 (Mac)/Ctrl~4 (PC).

With Pro Tools in Slip mode, any region you grab with the Grabber tool can be moved freely to any point in time on its own track, or even to any other track you may want to move it to.

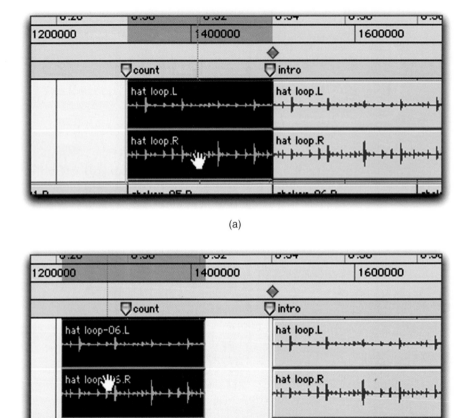

(a)

(b)

Figure 2.43 When in Slip mode, you can use the Grabber tool to drag regions to any point in time, or even to other tracks in the session.

Before you skip ahead to the next section in a fit of know-it-all boredom, take note that if you want to move a region from one track to the next, holding down your **Control (Mac)/Start (PC)** key will keep it from moving forwards or backwards in time (this will come in very useful when we're learning to comp vocals).

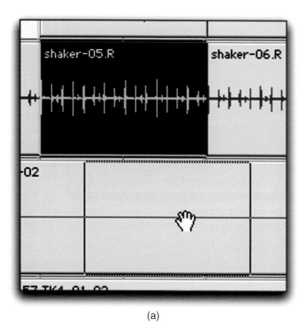

(a)

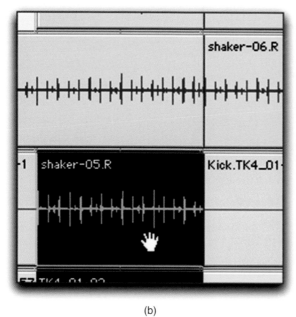

(b)

Figure 2.44
When dragging a region to another track (A) it can be easy to accidentally move it forwards or backwards in time. (B) To drag it to the track below and have it maintain its correct start and end time, hold down the Ctrl/Start key before selecting it.

To the left of the Grabber is the Selector tool. This is used to click and drag over anything from a short selection of audio you want to delete, to a snare drum that you want to zoom in on. You can think of this just like your favorite word processor – it just selects something that you want to do something to. To demonstrate, let's compare selecting and deleting audio with both Slip and Shuffle modes.

Figure 2.45 The Selector does just what its name implies. Shortcut key Cmd~3 (Mac)/Ctrl~3 (PC).

While still in Slip mode, click once and drag over a piece of audio like a bit of the shaker track at about 1.15 in the song, then hit the Delete key on your keyboard. You should see something like Figures 2.46 and 2.47.

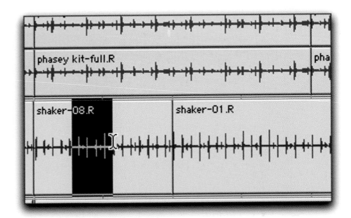

Figure 2.46 Section of audio selected with the Selector tool.

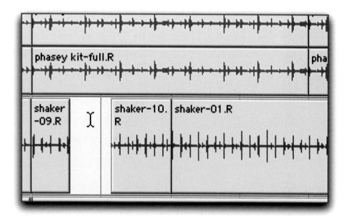

Figure 2.47   Now deleted. Notice the hole left in the audio.

Shuffle mode

Now we're going to try it with Shuffle mode turned on. Hit Cmd-Z (Mac) or Ctrl-Z (PC) to undo the edit you just made, then select **Shuffle** mode.

Figure 2.48   Shuffle mode will 'shuffle' the audio around. Shortcut key F2.

Make the same selection you made previously, and delete it. You should see the result in Figure 2.49.

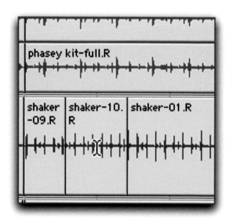

Figure 2.49   The remaining audio pieces 'shuffle' to the left to fill in the gap.

Shuffle mode 'shuffles' the audio pieces to the right of the deleted section to their left until they fill in the gap. The main problem with this is that it has now moved the entire rest of the shaker track out of time, so it is no longer with the beat of the song!

For this reason I usually recommend that beginners steer clear of Shuffle mode until they're reasonably comfortable with the workings of Pro Tools. Imagine if you were going through editing a drum track, with a low level of Undos in your preferences, and you edited several bars before you listened back. Every time you made a cut you would shift all of the audio to the right of the cut to fill in the gap. The track would become increasingly out of time and tempo, and you may find yourself without enough Undos to back out of it.

For now though, just Undo the edit and get back to Slip mode where we belong.

Next we're going to look at the Trimmer tool. Select it and you'll find that, just like the other tools (Grabber, Selector, etc.), it does exactly what it says.

Figure 2.50 The Trimmer tool. Shortcut key Cmd~2 (Mac)/Ctrl~2 (PC).

Hover your mouse over the right or left edge of one of the shaker regions and you should see it turn into the small bracket pointing away from the closest region edge. Clicking and dragging towards the center of the region will trim off the audio at the edge of the region, while dragging away from the center of the region will reveal any audio hidden underneath.

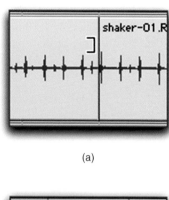
(a)

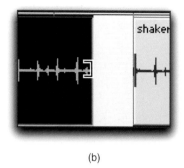
(b)

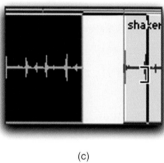
(c)

Figure 2.51 The Trimmer lets you trim off the edges of a region, or reveal previously trimmed audio that is still hiding beneath.

Grid mode

Let's move on to **Grid** mode, since now is as good a time as any. Click the Grid mode button and perform the same action with the Trimmer that you just did in Slip mode. Before you do that, though, change your Grid value box to Bars:Beats, 1/8 note.

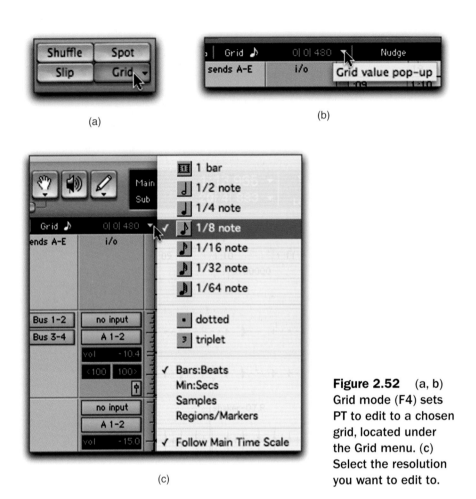

(a)

(b)

(c)

Figure 2.52 (a, b) Grid mode (F4) sets PT to edit to a chosen grid, located under the Grid menu. (c) Select the resolution you want to edit to.

As you use the Trimmer to edit the same shaker region, you should see that it now snaps to the resolution we set in the grid value, or every eighth note. This is only possible since we tracked this song to a click, which is a huge help when you get to the editing stage of a project.

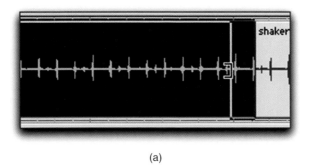

(a)

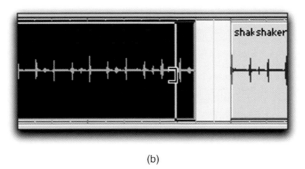

(b)

Figure 2.53 The Trimmer tool in Grid mode, editing the music to musical bars and beats.

Since the topic fits here, let me mention that you should always make sure your artists record to a click track of some sort. Whether it is the simple Pro Tools-generated click sounds (which are not recommended, as they tend to be real vibe killers) or a loop that has been time stretched to fit the Pro Tools session tempo, the time it will save you when editing is incalculable.

In terms of the 'feel' for musicians, a percussion loop feels better than a sidestick or cowbell to play along with. In a Nashville recording session, this usually involves the engineer sending MIDI timecode from Pro Tools so it follows the session tempo. This timecode goes out to the session drummer's own drum machine or click device, which the band then plays along to. However you do it, only if the session is recorded to the Pro Tools click will you be able to effectively use Grid mode, so whatever it takes to get there is worth it.

Moving along at the speed of light, we can summarize the Zoom tool with four simple words. I don't use it. While you could consider that five words with a contraction, I think you get my drift. Instead of the Zoom tool, let's talk about my approach to zooming in Pro Tools in general.

Figure 2.54 The lonely unused Zoomer tool – maybe if it didn't have such a dumb name.

Zooming

Zoom in/out

You probably know this already, but if you're zooming in and out on your tracks using any method but the Cmd-[ ] (Mac)/Ctrl-[ ] (PC) shortcuts, you're doing it wrong. This method will always zoom in and out using your timeline cursor as its center point, which is a major time saver while editing. While watching the Multi Platinum Pro Tools DVD, you will often hear a flurry of keyboard strokes as I zoom down on a note to make a quick edit or tune a note. This is Cmd-[ ], every time.

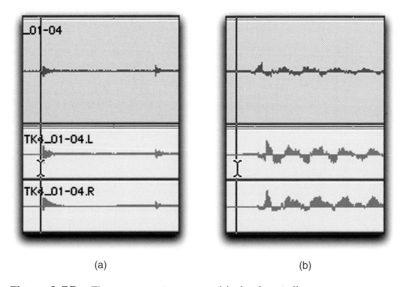

(a)   (b)

Figure 2.55 The same note zoomed in horizontally.

While there may be some great engineers that use another method, let's just call a spade a spade, and say that this is the way everyone should zoom horizontally.

What Digidesign calls **Vertical Zoom** is more often referred to as waveform size. This is useful when editing tracks that might not have been recorded at a sufficient level, or if you need to increase the relative size of the waveform to see difficult edits that need to be made. Option-Cmd-[or] will let you see everything you need to see.

Take the same waveform that we just zoomed in on horizontally. Notice that vertically speaking the drummer has not yet really kicked the volume up and the waveform is pretty small. Now, if we wanted to edit the top note we might have a difficult time, so we want to increase the waveform size or zoom vertically. A quick stroke of the shortcut keys should yield a result that looks like Figure 2.56B.

A quick note: As you do this, it may appear to the inexperienced that the waveforms are getting louder, simply because they're getting bigger. This is an understandable assumption, but with the zoom we're simply increasing the viewing size of the waveform, not its actual amplitude. If you consider a horizontal zoom in the same fashion, it appears that the waveform is getting longer the more you zoom in, but in fact you are just viewing it more closely.

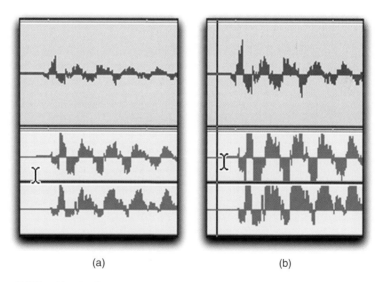

(a)   (b)

Figure 2.56 Vertical zoom increases the visible size of your waveform for ease of editing, but does not increase the volume of the track.

While not technically a waveform zoom, if you use a lot of other editing DAWs, you'd probably note that there's a third type of zoom to mention.

Unlike some other DAWs that group the Track Horizontal and Track Vertical zoom functions together, Track Height Zoom in Pro Tools is quickly accessed with the Ctrl-Up or Down Arrow (Mac)/Start-Up or Down Arrow (PC) keys on your keyboard. If you have placed your cursor on a particular track, it will just affect that track. If your cursor/timeline is across all tracks, they will all resize together. While not as elegant as some DAWs, I'm sure Digi will get their act together on this issue some day.

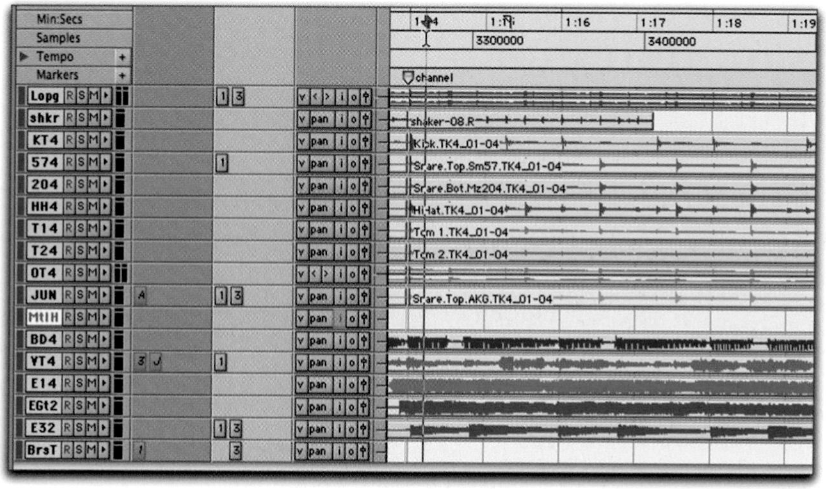

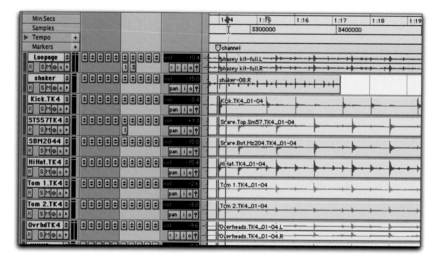

Figure 2.57 A group of tracks set to Track Height 'mini', with a Ctrl-Up Arrow (Mac)/Start-Up Arrow (PC) stroke turns into that shown in Figure 2.58.

Figure 2.58 All tracks increase their height to 'small', 'medium', 'large', etc. Larger track heights are more useful for editing.

Finally, let's look at two quick zoom functions that we'll be using all the time as we edit, the Zoom Out Full and Zoom In Full.

As we are working in a track, whether we're trying to find an errant bass note or nip a mouth noise from a vocal, it is very common for us to want to make a selection around a given note, and then zoom in super close on both a horizontal and vertical axis for micro-editing. This is done with a combination of a preference and a shortcut key.

To demonstrate, select any of the little shaker hits on the shaker track that you want to take a closer look at.

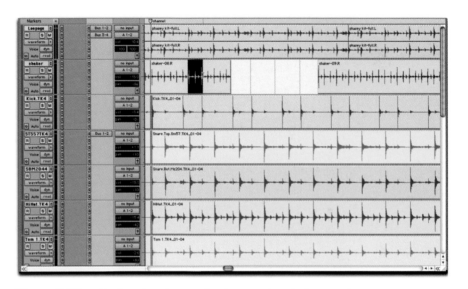

Figure 2.59   Notice the short selection on the shaker track.

Now, use the keyboard shortcut Ctrl-E (Mac)/Start-E (PC) to instigate the **Zoom Toggle** function.

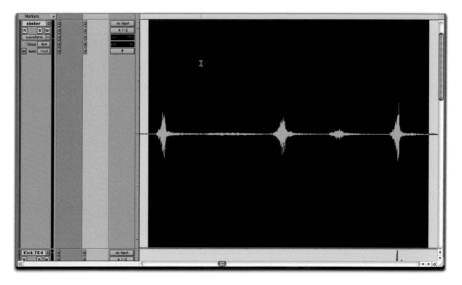

Figure 2.60 Zoom Toggle allows you to quickly zoom both horizontally and vertically on any selection.

Also located under the Zoomer tool, we will use Zoom Toggle on a very regular basis for quickly resizing all tracks around the track and note that we're currently editing.

Figure 2.61 Manually engaging the Zoom Toggle. Ctrl~E (Mac)/Start~E (PC).

If your selection filled the entire screen, like with the track size set to **Extreme**, we generally won't want the Zoom Toggle to blow the selection up quite that much. No worries, we can easily reset it to allow a few other reference tracks to still fit into our screen. To do this, simply engage Zoom Toggle over a particular selection, then resize the track to the height you want using the height ruler that we've explored before.

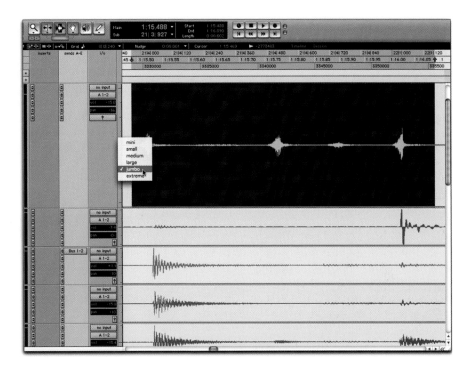

Figure 2.62 With Zoom Toggle engaged, change the selected track height to either Jumbo or Large depending on the size of your monitor.

By setting it to the more reasonable Large or even Jumbo track height, we're still able to visually reference the other tracks that we will be pocketing against. This is good.

Now, obviously you can zoom in and out of Zoom Toggle by using its shortcut repeatedly, but what about when you find yourself zoomed way in on a particular track, and all of your tracks are at different heights and sizes. When editing, you'll often find yourself in this position, and sometimes you just need to back out and see the entire song for a minute to get your bearings. To do this, we will combine one click and two shortcuts.

First, using your Selector tool, click in the timeline rulers directly above your audio tracks. This will place the timeline insertion selector across all tracks in the project. Then, repeatedly hit the Ctrl-Up Arrow shortcut to shrink all tracks down to a Small or Mini size. Finally, utilize your Option-A command to zoom out to view the entire session on the timeline. Check out Figures 2.63 and 2.64 for a visual.

Voila! In a few quick keystrokes you have gone from deep under water to soaring high over your session, and the timeline is still right where you left off editing, so you can quickly zoom in again once your mind is right.

Check out more on the DVD for some more zooming tips and tricks!

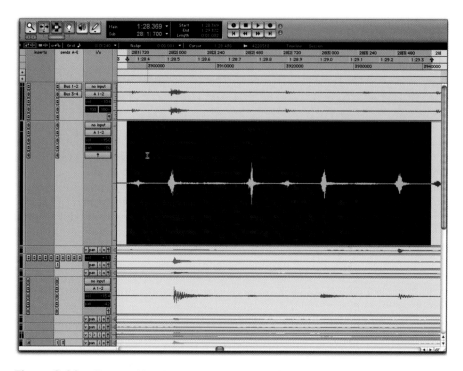

Figure 2.63 Zoomed in close with track at different heights, as in the middle of an edit. We need to get back to a quick overview of the whole song.

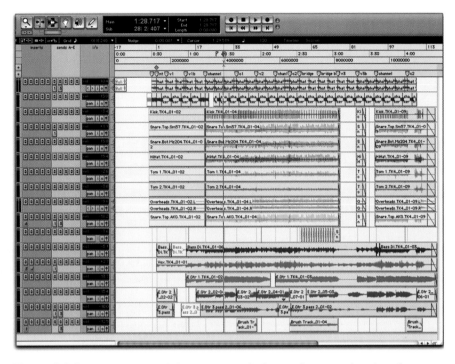

Figure 2.64 From super tight zoom to whole session overview in a few short seconds.

Shortcuts

Let's talk about my take on shortcuts for a minute. Keyboard shortcuts are great. They really are. And Pro Tools, as with all DAWs, has a shortcut key for just about everything you will do on a regular basis. For beginning engineers, one of the quickest ways to make yourself, well, quick, is to open up the Keyboard Shortcuts PDF under the Help menu and force yourself into an old-fashioned woodshedding mode.

Figure 2.65 Keyboard Shortcuts.

Open a session and practice every keyboard shortcut in the list from Apply Fade to Zoom Toggle, over and over and over. Thereafter, every time you find yourself using the mouse to do something you feel may be accessible with a shortcut, take 5 minutes and see if you can find it in the documentation. While you may be able to become lightning fast through sheer force of will (hey, there are fast typists who use the old 'hunt and peck' method), knowing the shortcuts will give you a huge leg up on the competition, and enable you to work both faster and smarter.

Since the topic of Mac versus PC shortcuts is bound to come up, let's summarize it by saying that any time you see a Mac or PC shortcut, or if you find yourself working on the opposite operating system in a foreign studio, know that almost every time your shortcut keys will translate as follows (Mac – PC): Option = Alt, Command = Control, Control (Mac) = Start button (PC), Control-click (Mac) = Right mouse click (PC).

The Smart tool

No discussion of Pro Tools is complete without mentioning the Smart tool. While we assume throughout the text that you are already coming in with a minimum level of Pro Tools experience, I find a lot of new users don't know much about the Smart tool. Introduced several versions ago, the Smart tool is a context-sensitive tool that changes depending on where our cursor is positioned over the waveform. Over the top half of a region it will become the Selector tool. Over the bottom it becomes the Grabber. On an edge it becomes the Fade, Trim, or Crossfade tool when hovered over the top, center, and bottom respectively. See the figures below for views of each one.

Figure 2.66   The Smart tool.
Shortcut key Cmd~7 (Mac)/
Ctrl~7 (PC).

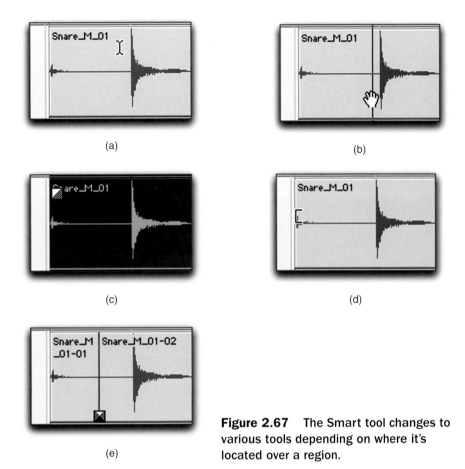

(a)

(b)

(c)

(d)

(e)

Figure 2.67 The Smart tool changes to various tools depending on where it's located over a region.

I've heard lots of engineers give their opinions as to why they do or don't like the Smart tool, but I personally use it all the time. In fact, every time we mention the Trim, Insert, Grabber, Fade, or Crossfade tools in this entire text, we're always talking about using the Smart tool version of them.

Finally, we will regularly be turning our groups on and off to check individual tracks or to edit all of a group together (like the drums, for example). Enable a group by selecting multiple tracks and hitting Cmd~G (Mac)/ Ctrl~G (PC). To suspend or reactivate a group use Cmd~Shift~G (Mac)/ Ctrl~Shift~G (PC).

Now that we have our session preferences set the way we like them, let's move on to analyzing and seeing how we set up the session we're going to be working with.

Beginning the pocket: Building a song from the drums up

Opening the session

Now that your session is set up, our preferences checked, and the screen laid out to our very specific liking, we're ready to work our way through editing this track together. You ready? Make sure you open the session titled Home_DVD Tracking.ptf.

Figure 3.1   Starting the edit with the Home Tracking session.

This is a song called *Home* by the artist Eric Paslay. He's an incredible singer/songwriter in the Nashville area, and this is a demo that we've recorded recently for shopping to record labels. Check him out at www.ericpaslay.com.

What you're looking at when you first open the session is just the initial bed tracks that were cut to give the song a starting place, before we get into a lot of the later production elements. Virtually nothing has been done to the tracks at this point, so like we've talked about, we're going start out by working our way through editing and pocketing all of the instruments. After completing the instrument pocket, we're going to start pocketing, comping, and tuning all of the lead vocals, as well as spend a bit of time with some serious background vocal editing.

The first thing we're going to work through editing will be the drums. Depending on the session, sometimes we may start with pocketing percussion elements first, but in this particular case, the drums need to be pocketed against the loop they were recorded against, before we work our way up through the bass, rhythmic, and melodic instruments.

On the night this was recorded, rather than using a metronomic click track, we chose to use a loop track that was aligned to the Pro Tools tempo.

Figure 3.2 All of the bed tracks available to us during our initial edit.

While we won't be going into the recording process in this text, I personally almost always use an application called Stylus RMX for the creation of great feeling, and very contemporary sounding groove loops for the artist and musicians to play to. It feels so much better to record to a great loop than to a bleeping click track, and your performances will be light years ahead when you utilize it. Our shaker track also came via Stylus.

Beyond these initial loop percussion elements, this track comes complete with drums, electric bass, some vocals, as well as electric and acoustic guitars.

Importing tracks

When you are working through a modern major-label multitrack production, while the initial tracks may still be recorded with a group of musicians located in a room, the process of overdubbing has grown well beyond these boundaries. As we'll explore later, an initial tracking session may be rough mixed down to an MP3, and emailed out to a great acoustic guitar player in New York. That player will then commonly reconvert the rough mix MP3 into a '.wav' file, import it into a Pro Tools session, record their guitar parts, and then upload the whole track to an FTP server for the producer and editor to grab. This new guitar track is then imported into the original session, along with all of its punch-ins, alternate takes, and metadata that the producer may be looking to use.

Like we mentioned at the beginning of the book, here's a time we want to show you a handy tool we used, but that you won't need to do in the included session. If you already know about inporting tracks and want to skip ahead, jump to page 75.

In this case, the artist played all his own acoustic guitars, but we recorded them in a different session. To import this guitar track into our current session for editing, we went to the File > Import > Session Data menu.

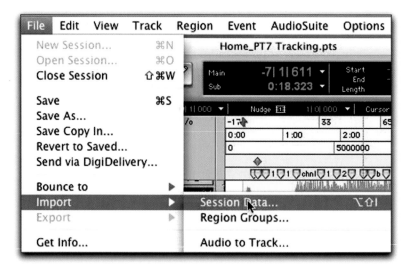

Figure 3.3 Importing session data.

We navigated through the Multi Platinum Pro Tools folder through the Home_Master to the Old Sessions folder. Selected Home_AcousticODS and clicked Open.

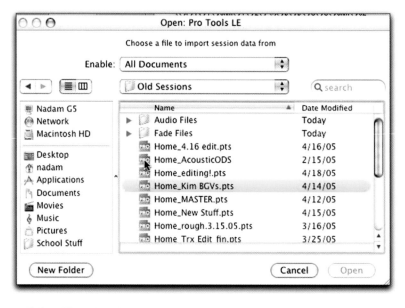

Figure 3.4 Choosing the session you wish to import from.

The following window came up, showing all of the information about the Home_AcousticODS session, including a list of tracks located within it. We dragged the scrollbar down to the bottom of the list and imported the Acoustic.TK4 track to a New Track.

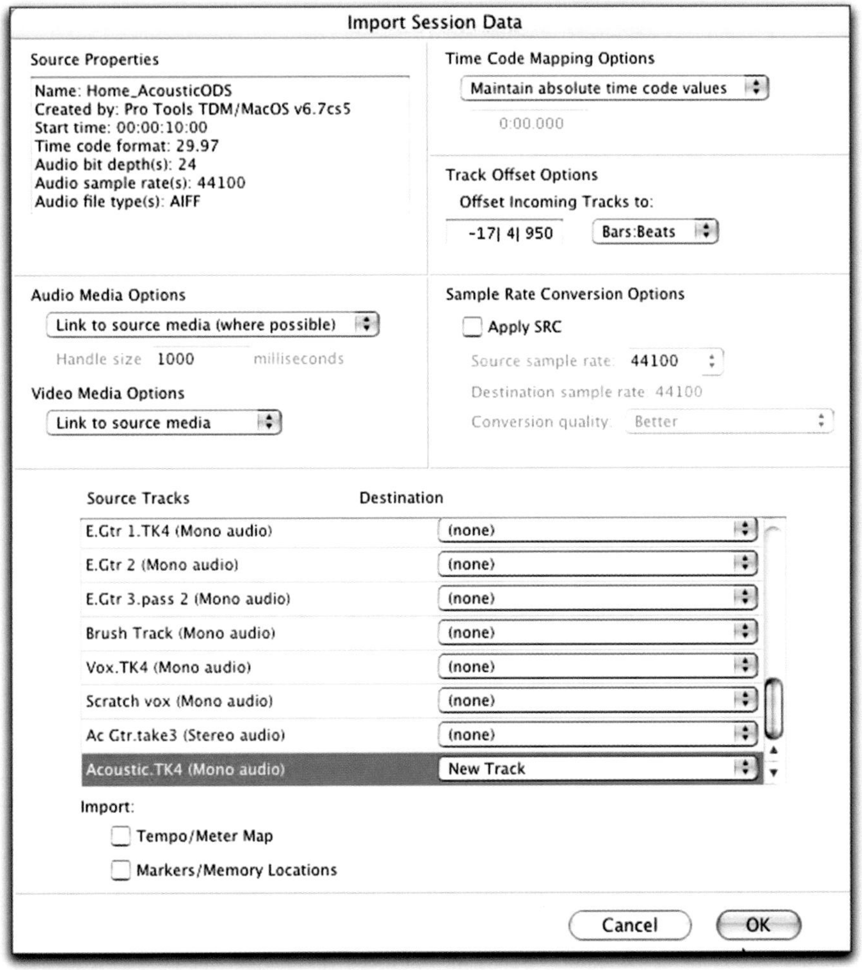

Figure 3.5   Importing acoustic guitar to a new track in your edit session.

You may find that when it imports the track it makes it hidden by default. We chose it in the Show/Hide tracks box and moved it to the very bottom of the window.

Figure 3.6   Showing the imported acoustic in the Show/Hide window.

You'll notice that the Acoustic.TK4 track is italicized in this window, and the track itself is grayed out. This indicates that the track has been made inactive, so obviously we need to reactivate it.

Activating the track

Since you may be using an older version of Pro Tools, in version 7 Digidesign made some pretty radical changes to the menu structures inside the software. Many functions that relate to individual tracks were previously found under the File menu. In version 7 they have been relocated to have their own new menu called, appropriately, the Track menu. We selected the acoustic guitar track by clicking on its name and chosing 'Make Active' under the Track menu to finish adding the acoustic to this session.

Figure 3.7   Activating the acoustic guitar track.

Finally, we needed to unmute the acoustic guitar and make sure its outputs are assigned with the rest of the track. You can also set its automation parameter to 'off' if you want to be able to mix its volume to your taste without having to hear any automation that's already been written.

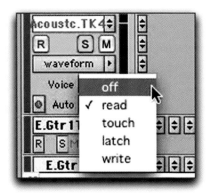

Figure 3.8 Disengaging automation on the acoustic track so we can hear it in the mix.

Taking stock of the session

Now, let's take a minute to take stock of what's in the session and what needs to be done. I currently spend most of my time editing tracks that I have been involved with since the initial pre-production meetings, all the way through the tracking and overdubs. As a result, I'm pretty familiar with what sorts of tracks we'll be working with. However, another large chunk of my time is spent editing, tuning, or pocketing tracks that I wasn't involved in recording, so it's always helpful to start off by looking over the whole session from top to bottom. Zoom all the way out with the Option-A command and scroll down through the tracks so you have a visual overview of what we'll be working with.

Top to bottom, you'll see the initial stylus loops, drums, a percussive brush track, bass, electric, acoustic guitars, and vocals.

What to view and how to view it: Cleaning up the Edit window

For any pocketing session I have a motto about what to view and how to view it. It goes something along the lines of, if I'm not working with it, I don't need to see it. That said, since we're starting out by pocketing our drums, we're going to Make Inactive or Hide any tracks that we don't need for the pocket we're working on at this exact minute. Start by selecting both the scratch vocal track and acoustic track, and selecting Make Inactive from the Track menu. This will free up some system resources for those on slower machines, as well as just getting them out of our hair. After you've made these inactive, go ahead and temporarily mute the electrics and bass guitars. We'll be getting to these tracks as soon as we're done with the drums, but for now we neither need to hear them nor have them cluttering up our workspace, so in addition to muting them, click to Hide all of the electrics, bass, acoustic, and vocals in the Tracks Show/Hide window.

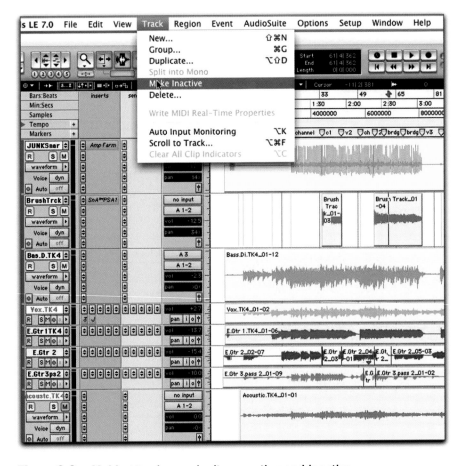

Figure 3.9   Making tracks we don't currently need inactive.

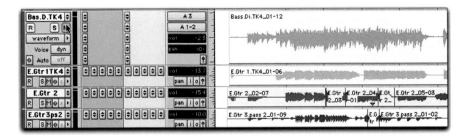

Figure 3.10 Muting the bass and electric guitars. We'll get to them later.

Figure 3.11 Hiding everything we're not editing to avoid clutter.

Now notice that we've got a brush track sitting below the drum kit. This brush snare was played by the drummer after the fact. At this point in the edit, I'm not entirely sure whether we're going to use it in the final mix, but it's here if we want to. You will find that when working with quality session players, often they will give you more great material than you can ultimately fit into the mix, so it's OK to have options that we may not end up using. Give it a listen real quick, and you may find you want to keep it in the mix for now. Personally, I'm going to mute and hide it for now with all the rest.

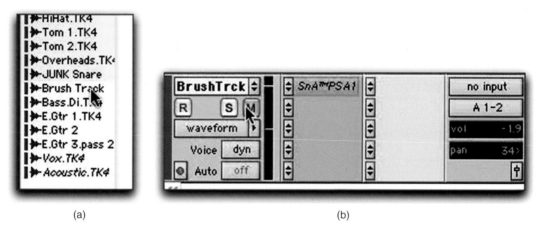

(a)

(b)

Figure 3.12   Sorry brush track. You are the weakest link. Goodbye.

Start by soloing and turning up the loop and the shaker tracks, since those were what we originally tracked the drums to, rather than a click. These will give you the feel that we're going to be pocketing against as we work through these drums, so it's good to give them a listen in solo to get used to the vibe they put out.

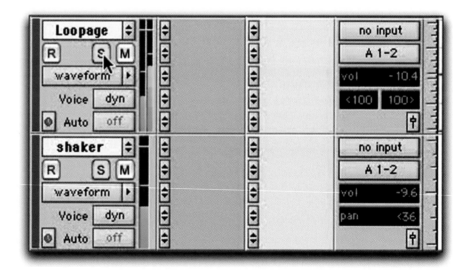

Figure 3.13   Soloing the loop tracks to hear the feel we'll be using for this song.

Now cut over to around bar 21 and solo the drums against the loops and shaker so you can see how they feel.

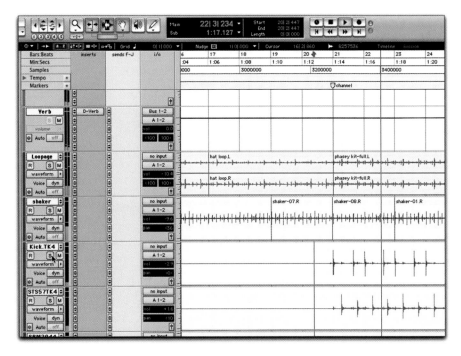

Figure 3.14 Soloing the drums against the loop.

Notice that there's just a bit of flamming happening between the snare and the loopage track. The first question to ask when pocketing any track is 'What am I going to line this up to?' Will you be using a click? A percussion element? The loop? Another instrument?

Obviously your ears need to be the be all and end all judge when making this determination of what sounds good and what doesn't, but for finding a visual guide track to work against, we would prefer something that has a sharp, defined transient, with enough level to visually pocket off of.

Finding a visual guide track

To that end, the shaker will probably not be our best friend in this particular track. For example, zoom in on the shaker transient located about bar 20, beat 4. The transient of a shaker is really long and undefined, making it very difficult to determine exactly where any kind of attack happens, let alone where the start and end of the beat is.

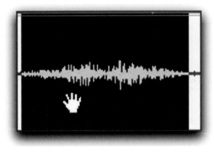

Figure 3.15 A shaker transient – unsuitable for pocketing against.

Our best bet is going to be to use the hi-hat-based loop track, with a little bit of vertical zoom to accentuate the transients of each hit. For a visual go to the loop transient right above your shaker, and use your vertical waveform shortcut to beef it up.

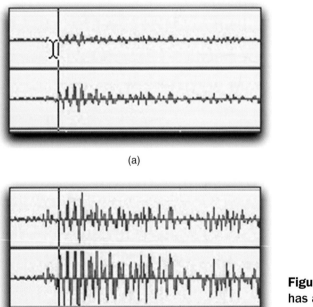

(a)

(b)

Figure 3.16 The loop has a much more defined transient than the shaker.

Since we've settled on the hat-loop as our guide track, let's mute and hide the shaker to further clean up our workspace.

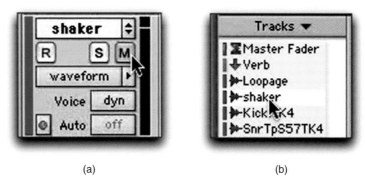

(a)   (b)

Figure 3.17   Losing the shaker.

We should now just be viewing our drums and our loop. Make sure your drum group is enabled, and temporarily set your drum track height to small.

(a)

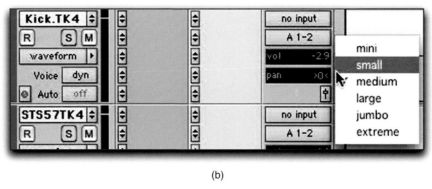

(b)

Figure 3.18   Engaging your drum group and viewing them all on the screen.

Editing within a drum group

We engaged the drum group so that every edit we make to any track on the kit will be applied to all members of the kit. If we just tried to edit the kick, snare, or any other member of the kit by themselves, the track would quickly fall apart because the sound of each drum is going to bleed into every other track of the kit as well. So, if we pocketed any of the tracks individually, say the snare track, you would hear both the new, tight, pocketed snare, and the now out-of-time bleed of the snare track in the overheads, tom, and hat tracks. Obviously, that won't sound good, so we're going to edit all of our drum tracks together.

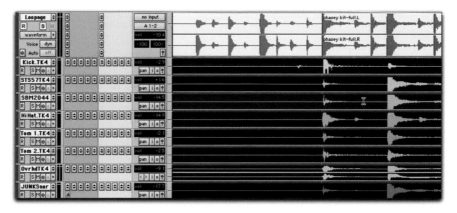

Figure 3.19 Selecting to make sure the group selects across all drum tracks before we start editing.

Now that we know every edit we make to any of the drum tracks will be carried across to all drum tracks, we really only need to view the tracks with the biggest transients for pocketing against our loop. This is going to be our kick and our snare tracks.

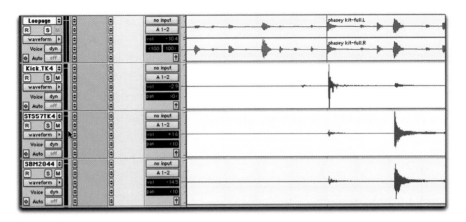

Figure 3.20 Using only our large transient tracks to pocket with.

Track-naming conventions

If you're thinking that the track labels/names look a little funny in these fig-ures, you can always open them up and change them to something you will recognize more easily. I tend to use at least two snare mics, and as a result like to put the microphone used as well as the final take we end up using in the track name. It's a personal convention that I've found useful. I named the tracks in your session a little more plainly for any Pro Tools newcomers following along.

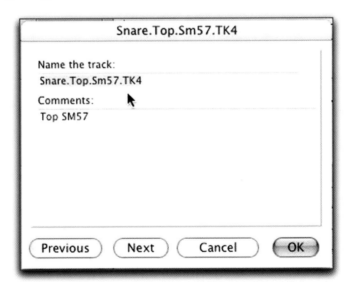

Figure 3.21 Putting a little extra info in the track name dialog box.

'So,' I hear someone ask, 'if you're not going to need to see anything but the kick, snare, and loop while you're pocketing drums, why don't you hide the rest of the drum tracks like you have all the other instruments?' Good question. The reasoning is simple. If you have a group of instruments that you're wanting to edit together (this applies to drums, background vocals, or any other ganged sounds), if I hide, for example, the hat track and make an edit across all of the other drums, as soon as I unhide the hat track, uh oh – no edit was made. Basically, any edits you make do not affect a hidden track, even if it's a member of a group.

Figure 3.22 Hiding the hat track that should be edited with the drum group.

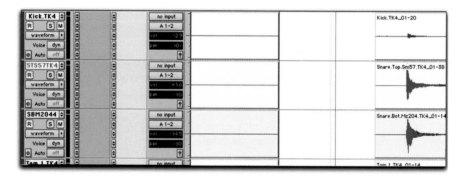

Figure 3.23 Making an edit to the remaining tracks of the drum group.

Figure 3.24 Unhiding the hat track reveals that …

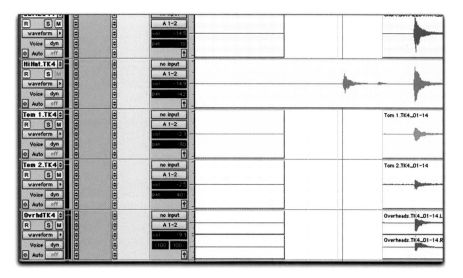

Figure 3.25   Ouch – the hat track was not edited with the rest of its group.

So, all that to say that you need to be viewing every member of a group that you're going to be currently editing – otherwise you'll make a lot of edits, and waste a lot of time before you realize you have to hopefully restart from an earlier version with intact drums that you can pocket all over again. Go ahead and undo your way out of that mistake.

To start the drum pocket, go back to your Show/Hide window, and drag your loop track down between the kick and the top snare track. This will give you a view that lets you see the loop with the kick above and snare below.

Figure 3.26   Moving the loop between the kick and snare.

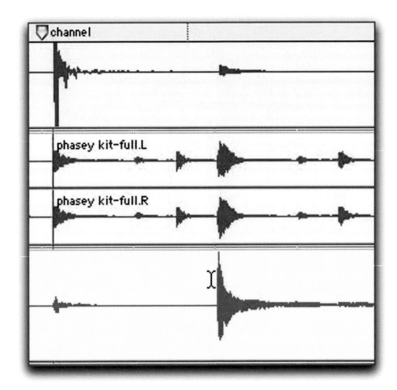

Figure 3.27   Your tracks should look like this for pocketing the drums.

Now with your drums grouped, this arrangement will allow us to edit our drums by selecting and editing in either the kick or the snare track, and our edits will of course be made to all of our drum tracks. The reason we put the loop between the kick and the snare is because we're going to zoom in very close to compare the transient of the kick drum with the transient of the loop. At the times where there is no kick present, we will be using the snare drum for comparing to the loop transient. By having the loop track in between them, we can easily and visually compare either the kick or the snare track to the adjacent loop track, without having to visually skip over another track in the process.

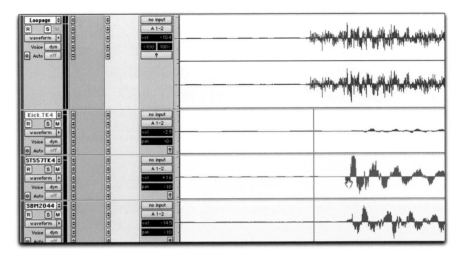

Figure 3.28 This is the wrong way to arrange the tracks. Notice that if we want to compare the snare transients with the loop, we have to try to look over the kick track.

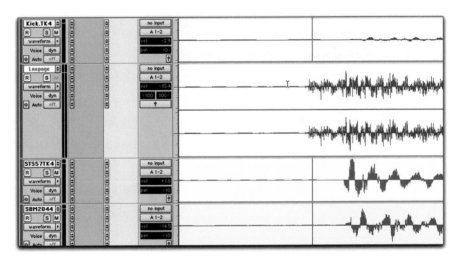

Figure 3.29 Now the snare transients and/or kick transients will both be adjacent to our loop transient for pocketing purposes.

What to do when there is no kick or snare

What do we do when there is neither a kick or a snare for pocketing against the loop? I tend to use the drum track which has the next largest transient, which is often the hi-hat. I rarely use the bottom snare because it is often farther away from the snare than the top mic is. As a result there is commonly a significant timing/phase difference between it and the top snare mic, so if we pocketed to the bottom snare, it would actually make our drums feel too on top of the beat, or rushed. If all else fails and it's a spot with no kick, snare or hat, like a tom fill, you can temporarily drag your tom tracks up to pocket against them. Ninety-five percent of the time we will be using just the kick and snare though, but go ahead and drag your hat track up beneath your top snare track to use as a backup.

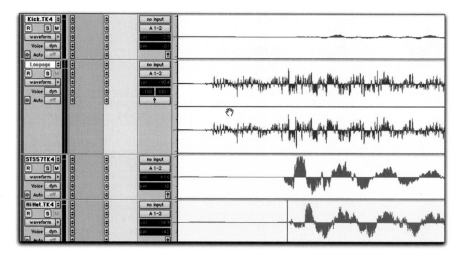

Figure 3.30 Now our kick, snare, and hat are surrounding our loop track for pocketing.

Excess noise cleanup

To begin to pocket, I like to just clean up my tracks by eliminating any excess bits of noise or fooling around. At the top of our song the drummer was doing a little warmup that we don't need, and I know this because we've already put our markers in the song showing our countoff and our intro. Anything before these tracks can almost definitely go.

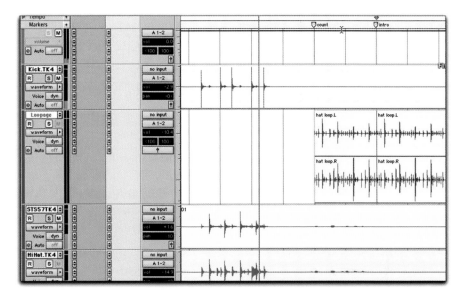

Figure 3.31 Excess noise at the top of the song. This needs to go.

This leaves us with a few options. After vertically zooming and listening through the countoff section to make sure the song doesn't have any gentle cymbal hits or creative drum intros, we can choose to either delete the excess or simply mute it. In the case of a drum countoff, as sure as we delete it the artist is going to want to come back in for a different ripping cool guitar part over the intro to the song, and we'll have deleted the countoff, making life very difficult. So for now, I just choose to make a cut before the drums by dropping in my timeline selector right before the drums come in and separating the regions by using Cmd-E (Mac)/Ctrl-E (PC). Now, just select the noise before they come in for real and use the Option-M (Mac)/Alt-M (PC) shortcut to mute the unneeded regions.

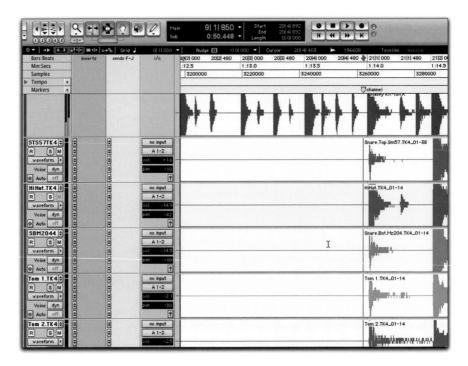

Figure 3.32 Separating the good from the bad with the Separate Regions shortcut.

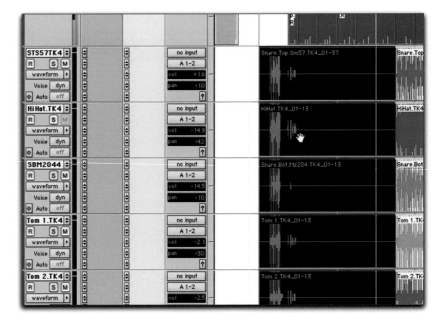

Figure 3.33 Using the Mute Regions command to just mute, rather than delete them. You never know when you might need that countoff again.

Now let's trim up the excess at the front of the first drum hit and use our Smart tool to put a quick fade across the drums around bar 21. Zoom in a little closer if you need to. Be sure to not let the fade go long over the actual transient of the snare or you will dull down that initial strike, which will bring down the energy of the song. This fade is just to prevent any little 'pops' that may happen if there was some stray bit of sound near our edit. Quickly make sure you're in Slip mode, and after this, you're ready to go.

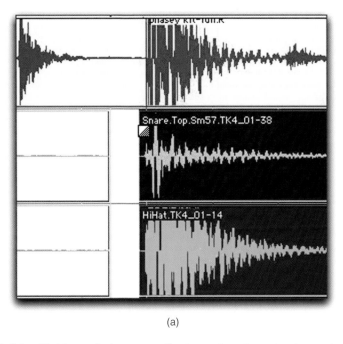

(a)

Figure 3.34 Making a fade across the incoming drum tracks, and not fading over the transients.

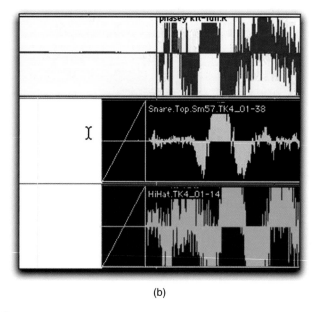

(b)

Figure 3.34 Incoming fade on the drums.

Pocketing our first note

When you sit down to pocket a track you need to decide exactly what part of the transient you are going to define as the downbeat of your drums. Kick drums can have all kinds of different shapes and styles to their transients. Some may choose to view the first hint of the waveform as their transient, some will choose to set the beginning of the earliest 'big' waveform as their transient. The fact is that, believe it or not, it really doesn't matter which part of the transient you choose to use as your pocketing guide, so long as you are consistent throughout the entire track.

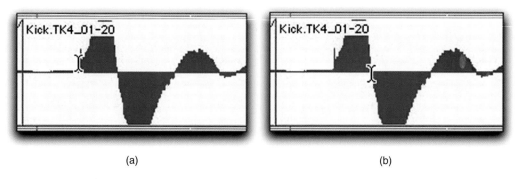

(a)   (b)

Figure 3.35 Which point in time will you call the beginning of the transient of your kick drum?

Whether we tracked this session to a click or a loop, most of the time that click is just being used as a timing guide, and it will not actually be heard in the finished mix. So if we define Figure 3.35B as the downbeat, and align that with the front of our click/loop, as long as we consistently use that part of the wave as our downbeat, it will work. It may sound like we're trying to push the song towards one feel or another, but consistency is the key to a good pocket.

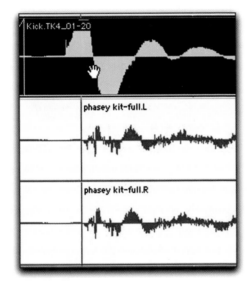

Figure 3.36   Choosing your transient and lining it up with the transient of the loop.

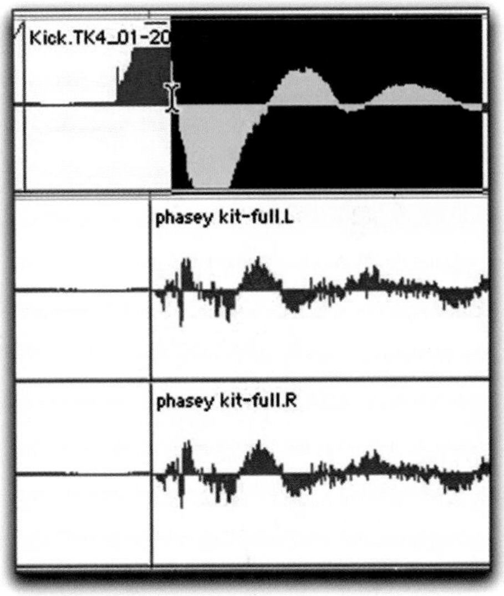

Figure 3.37   If you choose this part of the transient as the downbeat of your drum, you must consistently choose it as the downbeat of your drum.

For me, I'm going to use the front of this kick transient as my downbeat. So I'm going to grab it with my Smart tool and move it to a few milliseconds behind the front of the loop transient. While we may or may not ultimately leave this loop in the mix, putting the drums just a few milliseconds back will help our track to feel a bit more relaxed, with a bit more vibe. Putting the drum transient ahead of the loop or click will add a bit of an apprehensive or anticipatory feel, which won't work for this song.

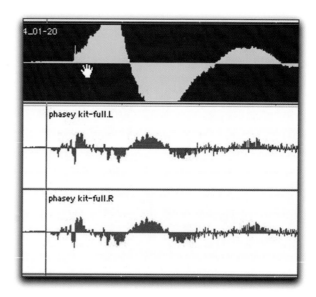

Figure 3.38 Moving the transient of the kick just a few milliseconds behind the transient of the loop.

If you're of the 'I need an exact number that I should be using!' crowd, change your main counter from Bars:Beats to Minutes:Seconds under the View > Main Counter menu, and make sure you've moved the front of the kick transient about 2–4 milliseconds behinds the front of the loop.

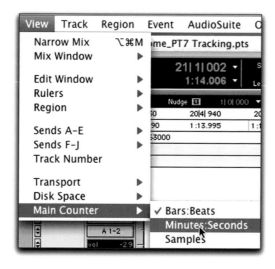

Figure 3.39 Changing your counter to view Minutes: Seconds.

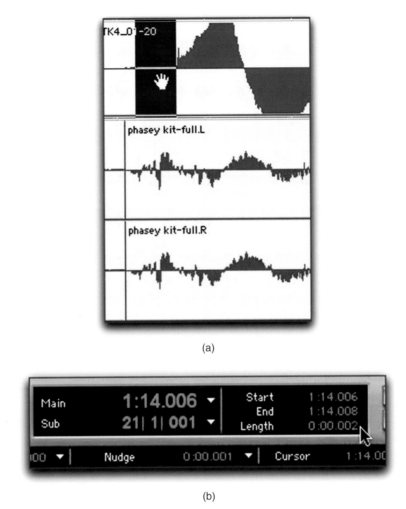

(a)

(b)

Figure 3.40 Make sure your kick transient is about 2–4 milliseconds behind the front of the loop.

Congratulations! You've just pocketed your first note! Only a few thousand more to go and you're going to be a certified master editor for sure! Now let's move on to edits 2–9999. If you need to, zoom out for a minute to get your bearings and find the next major transient. In this case it's going to be your snare hit.

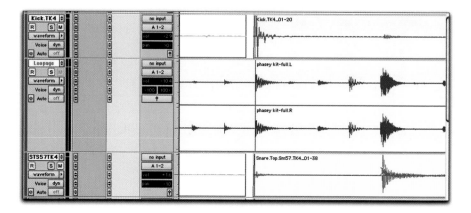

Figure 3.41   Notice the next major transient to pocket is the snare hit.

Moving on: A drum pocketing system

Now to start the repetition. At a medium/close level of zoom like you see in the figures below, we are going to drop our Selector tool in right before the transient, use the Separate Region command (or shortcut) to separate all of the drum group regions, then drag the regions back until the transient is sitting around 2–4 milliseconds back from the top of the closest loop/click. Check it out.

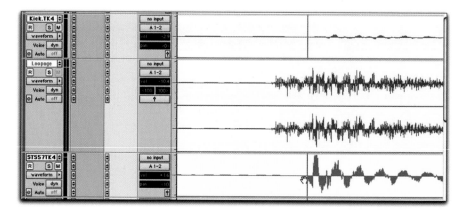

Figure 3.42   Drop the Selector insertion right before the snare transient.

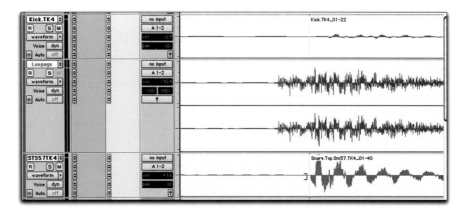

Figure 3.43   Separate the regions. The 'I' will turn into a Trim tool after separation.

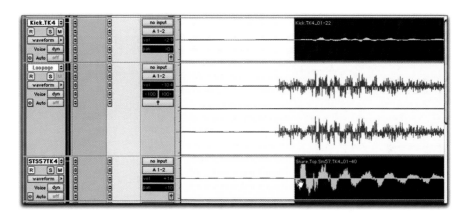

Figure 3.44   Drag the drum transient back until it's about 2–4 milliseconds behind the loop transient.

Now use your Trim tool to trim the edge of the snare region back away from the transient just a bit.

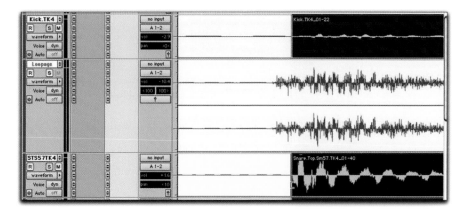

Figure 3.45   Using the Trim tool to add some space to the edge of the snare region.

And finally, hover your cursor between the two regions to create a small crossfade to clean up your edit. Be sure to not crossfade over any part of the transient!

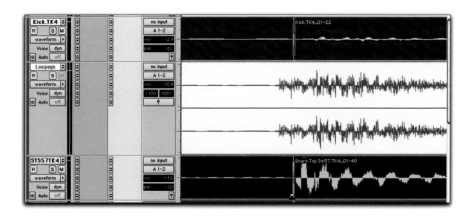

Figure 3.46   Making a tiny crossfade to cover the edit.

Now, repeat that process until the drum track is finished and voila! You've just completed the drum pocket.

OK, you're right, it's not quite that easy. There are a *lot* of special cases that require unique handling to make a good pocket. Now that you've mastered the base level skill, let's continue to move through the track together to see what sort of challenges it will present us.

Zoom out a bit and find your way over to the next kick transient, select right before it, zoom in, and separate the regions.

What to do when you have no transient to pocket to

Now that you're here, you may notice that there doesn't appear to be anything on the loop track to pocket this to. Such is the occasional problem with loops.

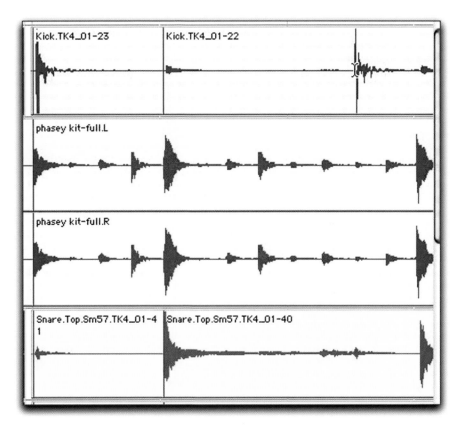

Figure 3.47 Zoomed out, and setting our Selector right before the next kick transient.

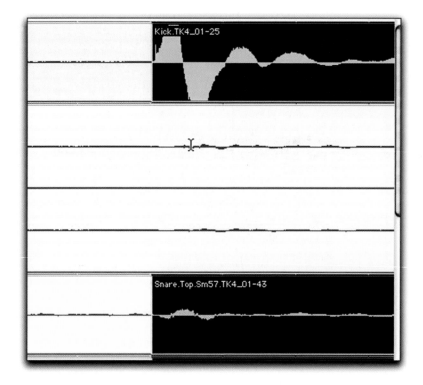

Figure 3.48 We've zoomed in and separated the region, but there's no loop to pocket against – or is there?

In this case, there actually is the barest hint of a loopy hi-hat sort of transient that begins right about where the cursor is above. We just need to see it better. To do this, go back to your old vertical zoom shortcut to magnify all the waveforms so you can see well enough to pocket the note.

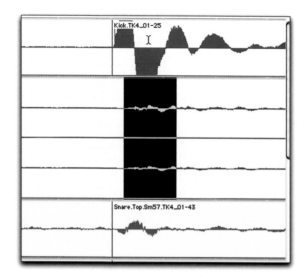

Figure 3.49 The vertical zoom magnifies the loop transient enough for us to pocket our kick drum against.

Now you can grab the kick, move it back, and pocket it just a few milliseconds behind the loop. Again, don't pocket the downbeat clear at the front of the loop or it will feel rushed. Set it back just a few milliseconds to maintain the vibe of the song.

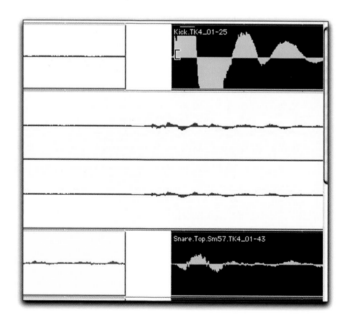

Figure 3.50 The kick pocketed a few milliseconds back.

After it's in place, repeat the process of trimming the audio back to fill the gap, and run into our next big problem.

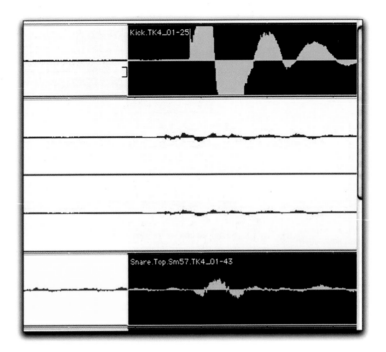

Figure 3.51   Fill in the gap created with the Trim tool.

Double transients in drum pocketing and how to deal with them

You may have noticed the problem if you jumped ahead and tried to do the crossfade. Shame on you. For those of you still following along, the problem we run into is this. Because we've had to move the audio so far back to bring it into time, if we try to drop in a crossfade we will have a double transient. Hmmm.

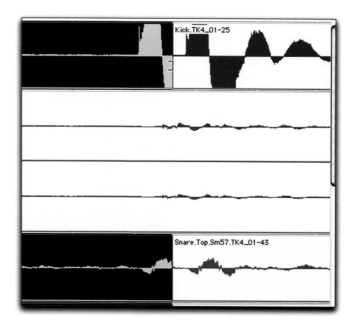

Figure 3.52 Moving the audio into the pocket would result in a double transient if we add a crossfade.

There are a couple of ways to handle this. Occasionally, we can just drag the edit back a little earlier in time and drop in a crossfade before the double transient would occur, like in Figure 3.53.

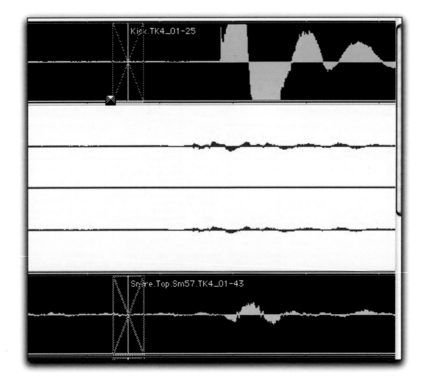

Figure 3.53 Just dragging the edit and crossfade earlier to avoid the double transient.

The problem with this workaround is that the audio to the immediate left and right of the crossfade is the same piece of audio! Often, if you try to get away with an edit like this, you will actually hear that repeat in the audio. This is, as we say in horribly mangled Spanish, no bueno. The smoother way to actually accomplish this edit is to have a quick discussion about the power of time compression/expansion.

Using time compression/expansion to fix a drum edit

If you jump up under the AudioSuite > Other menu you will find that Digidesign actually includes a Time Compression/Expansion tool inside of Pro Tools. As of version 5 (in TDM, 6 in LE) Digidesign has actually built the functionality of the tool into the Trim tool as a drop-down option.

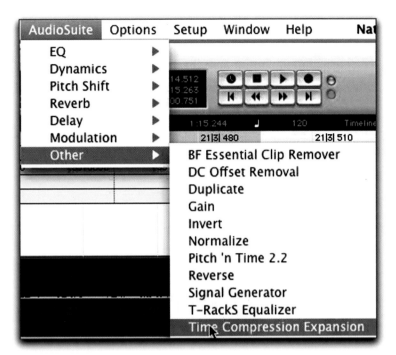

(a)

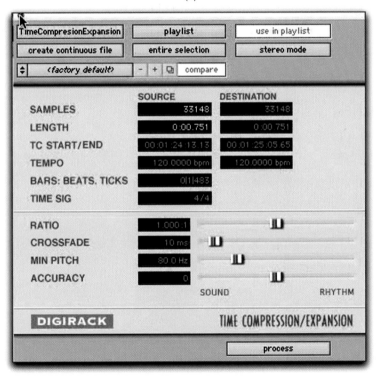

(b)

Figure 3.54 Digidesign's built-in TC/E tool.

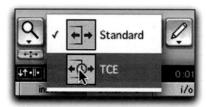

Figure 3.55 The TCE option under the Trim tool. Hit Cmd~2 (Mac)/Ctrl~2 (PC) twice to activate TCE.

The difference between Serato and Digi TCE tools

In addition to using the built-in Digi TCE, there are also a variety of third-party utilities that do essentially the same thing. From my experience, each has its own strengths and weaknesses.

While I feel that the Digidesign TCE tool tends to do a really good job time stretching vocals, my personal favorite for the stretching of instruments is a tool called Serato Pitch 'n Time.

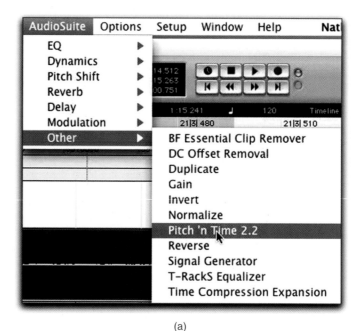

(a)

Figure 3.56 Serato Pitch 'n Time is a great alternative for instrument time compression/expansion.

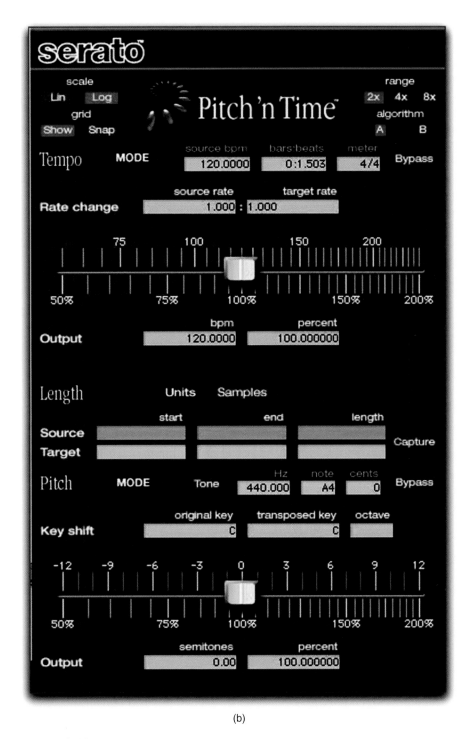

(b)

Figure 3.56 (Continued).

In order to map the Pitch 'n Time engine to the TCE Trim tool, let's revisit our old friend Preferences and look under the Processing tab. At the bottom of the Processing window you will see the options available for the TC/E plug-in. When clicked on, any of the compatible TCE plug-ins you have on your system should show up as checkable: Wave Mechanics Speed, TimeMod, Time Shifter, etc. All of these are great plug-ins, but if you don't have them installed on your system, they'll just be grayed out.

Note: I know not everyone reading this book will be able to run out and spend $600 on a copy of Pitch 'n Time (but really, who needs to pay rent this month?). If the default Digi TCE is your only option, have no fear. Everything we're going to do with the TCE tool will still work, it just might take a little more work to take care of any small audio glitches that might come up as a result.

For now, I'm going to go ahead and choose Pitch 'n Time and close out the Preferences window. When we run into some issues with time-stretching vocals, we'll alternate back to Digi's plug.

Figure 3.57   Switching the TCE default plug-in engine.

Looking back at our double transient problem, we need to fill in that hole left by our pocket. Note, though, that if you try to just time-stretch the drum region to fill the hole, you'll get a pleasant little error message saying you cannot perform that trim, because the region has a fade on the other end.

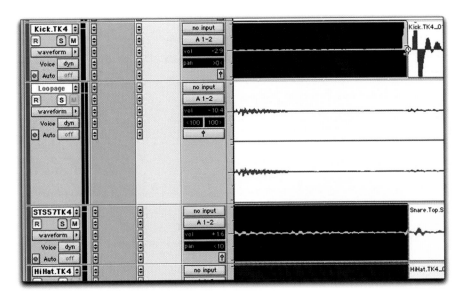

Figure 3.58 Attempting to use the TCE Trim tool to fill the hole.

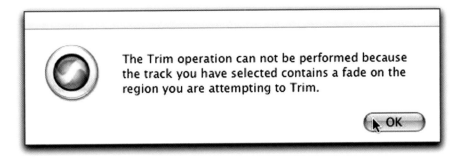

The Trim operation can not be performed because the track you have selected contains a fade on the region you are attempting to Trim.

OK

Figure 3.59 The resulting error message, because we put a crossfade on the front of that region already.

Hit the Option tab command to have Pro Tools jump to the previous edit on the drum tracks, and you will see the small crossfade we put on the beginning of that snare hit, and all the grouped drum tracks.

109

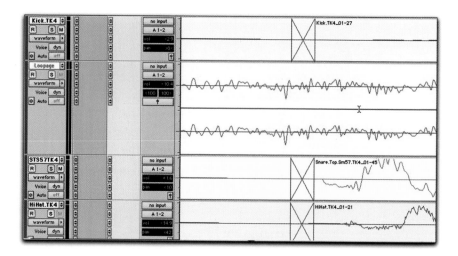

Figure 3.60   Sure enough, the crossfade we put in front of the snare edit.

Proper use of time stretching when fixing drums

Now, while an error message may be a compelling argument not to do something, the fade isn't the only reason I don't want to time-stretch that entire note. Think about it. If we time-stretch the entire region from this point to the next kick transient, all of the notes in that region will slow down quite audibly. The other reason is that the human ear/brain combination uses the transients of a sound to help identify what it is that we're hearing, so we want to keep our transients as pure as possible. If we prevent stretching just the front transients of our waveforms, it will help in making the parts we do stretch sound natural.

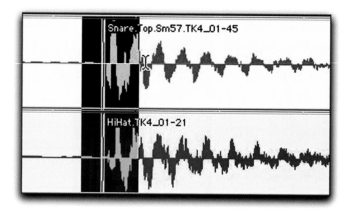

Figure 3.61   We don't want to time-stretch the transient of a sound – it will make the entire note sound unnatural.

So instead of stretching the entire note, let's go in and make a cut right after the main attack of the note.

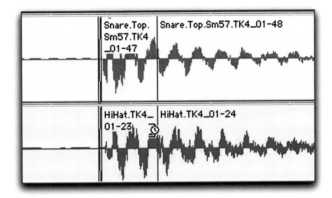

Figure 3.62 Drop your Selector tool right after the body of the note attack, and separate the regions with Cmd~E.

Now, assuming you have your Tab to Transients feature still turned on, tab back down to our next edit point with the double kick transient. Now zoom in a bit so you can see what you're doing, and use your TCE Trim tool to stretch that piece of the old transient until it fits over the new repetition of it exactly. Check out Figure 3.63 for a visual example.

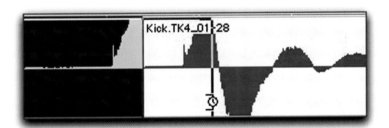

Figure 3.63 Note the piece of old transient being stretched to match its place on the new transient with the TCE tool.

Figure 3.64 When you let go, Pitch 'n Time (or Digi TC/E) will stretch the region to fit.

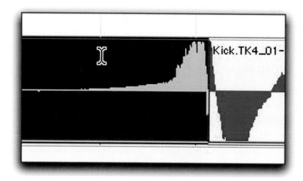

Figure 3.65 Voila! You've successfully stretched the region to match the next edit. Just one more step to make it sound right!

Now switch back over to the standard Trim tool, and you can trim back the region to before the correctly pocketed transient. Drop in a crossfade, then Option tab back to the splice you made right after the initial attack of the previous note. Finally, use your Smart tool to add a super tiny (say, 2 milliseconds) crossfade to cross between the unaffected transient, and it's now time-stretched back half. If you can't yet eyeball what a 2-millisecond crossfade looks like, look up under your 'Length' counter, which will better help you approximate.

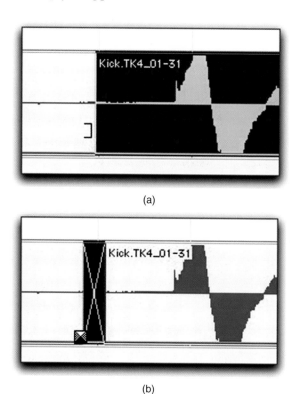

(a)

(b)

Figure 3.66 Trimming back before the correctly pocketed kick, and adding the crossfade.

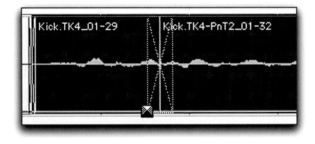

Figure 3.67 Add a quick crossfade between the unaffected transient and it's time-stretched back half, and your problems are solved!

Figure 3.68 The Length box tells you how long your crossfade is.

Adjusting the bounds to fit your fades

When you attempt this crossfade, you may get a message that says something to the affect of Pro Tools not having enough audio data to create one or more of the fades. This is because while the untimestretched region to the left of the fade has all its underlying audio data and could be trimmed back out to its original length, the region on the right has just been created by the time-stretching function, and as a result doesn't have any extra to participate in a full crossfade. Just tell the error message to Adjust Bounds to whatever it has enough audio data to complete, and go on your merry way. It will adjust the crossfade over to the right, where there is enough audio to combine the pre-fade audio with the time-stretched audio, the hole will be filled, the fade will be made, and if it's done with Pitch 'n Time, it will sound perfect!

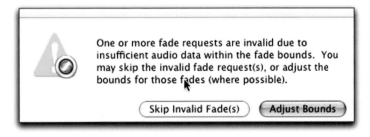

Figure 3.69 You'll get this message every time you try to fade between original audio and its time-stretched brethren. Just choose Adjust Bounds and it'll be OK.

Let's finish up with just a couple more notes. Repeat the procedure for the next snare located around 1.15. Tab to it, separate the regions, and drag it back to 3–4 milliseconds after the transient of the loop.

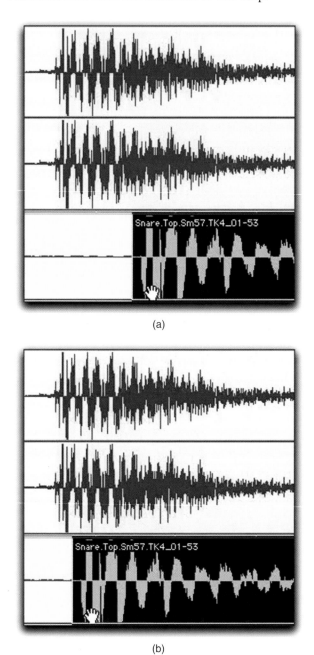

(a)

(b)

Figure 3.70   Tab to the next snare, separate it, drag it back.

While you may initially need to measure your gap with the Length counter and your Selector tool, you will eventually get a feel and an eye for what a 3- to 4-millisecond gap looks like.

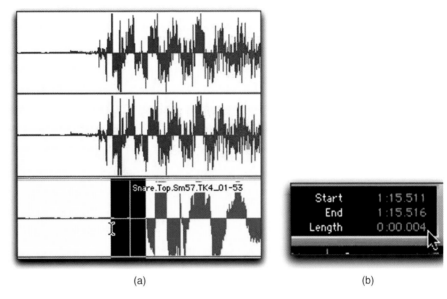

(a)   (b)

Figure 3.71   Using your Selector tool to measure out 4 milliseconds.

Once that's done, simply drop in your crossfade and move on.

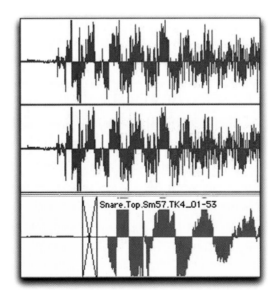

Figure 3.72   Drop in your crossfade and tab on to the next note.

While this may look easy, and really the mechanics obviously aren't hard, be aware that the temptation will be to just edit visually. Once you get used to chopping and moving those waveforms, it can quickly become almost robotic, and you can go for a long time while forgetting the single most important thing – *listening*!!!

The key to good drum editing

The key, when editing by both eye *and* ear, is to keep what the *ear* hears in a higher position than what the eye sees.

Make sure that you stop every couple of minutes and listen to the last few edits. Check for bumps, pops, or hiccups in the audio that need to be ironed out. To make this easy, go ahead and bring up your Transport window by hitting Cmd-1 (Mac)/Ctrl-I (PC).

Figure 3.73   **The Pro Tools Transport window.**

Using pre-roll to check your drum pocket

While this setting will change depending on what phase of the recording process I'm in, when editing I like to turn on about 5 seconds of pre-roll, so that Pro Tools always plays the last 5 seconds before the edit I'm currently working on. This helps keep your work in context and keep your bearings straight. Once you have your pre-roll time set, you can always turn it on or off with Cmd~K (Mac), Ctrl~K (PC) as well.

Figure 3.74   Setting pre-roll.

As you work your way through the next few notes you'll find a case where you need to use your hat track to pocket with, a few notes that will probably already be dead on where they need to be, a ghost snare or two that will need to be vertically magnified for visual alignment, and a myriad of notes and situations like the ones we've just gone over.

Pocket it or leave it alone? A rule of thumb

My general rule of thumb is that if I get to a note and it's no more than 4–10 milliseconds behind the beat, I'll leave it alone. For most music, we're not looking to quantize it to death, just to make it the best performance it can be, without robbing it of its humanity. If the note is in any way on top of the beat, though, I'll virtually always pocket it even slightly behind, just so it maintains that laid back feel.

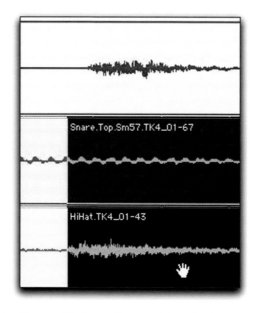

Figure 3.75   I will most likely pocket anything that's too on top of the beat or it will feel rushed.

Once you have worked your way through the next 20 or 30 seconds of audio, jump back to the beginning of the track and, starting with the loop, let it roll through the intro, where the pocketed drums come in, and on into the stuff yet to be pocketed. If you've been pocketing properly, you should immediately be able to hear how even a well-played drum track that felt good before pocketing has now been made even tighter, more punchy, and with an even better feel. If you've moved the unpocketed audio too far out of sync, use Spot mode to put it back for comparison purposes.

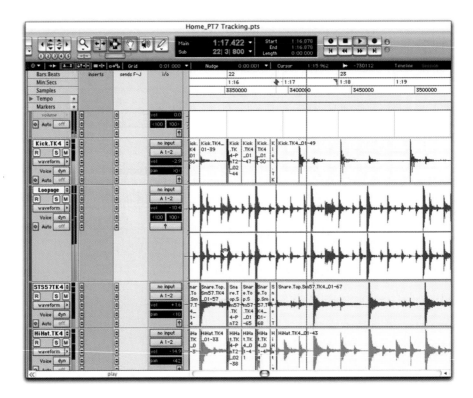

Figure 3.76 Pocketed versus non-pocketed drums. Do you feel the difference?

Using Spot mode to bail out when you lose your editing perspective

A quick word on bailing yourself out when you lose all perspective. Say you've been editing for quite some time, and find yourself moving the drum track significantly farther than you know it should be, but you've lost the sense of where it started out at.

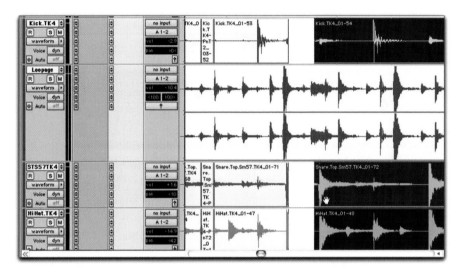

Figure 3.77   Where did these drums go again?

When this happens (and it will happen), we utilize that last mode that is so under-appreciated in Pro Tools, Spot mode.

Figure 3.78   Spot to the rescue. Spot mode = F3.

When in Spot mode, clicking on the audio will bring up a dialog box showing the current time stamp as well as the original time stamp of the selected audio. By clicking the arrow next to the Original Time Stamp, you can place the selected audio back at the original point in time it was recorded at – like a big giant Reset button.

Spot Dialog

Current Time Code: 0:0-10.00-9 ▼

Time Scale: Min:Secs ⬍

Start: 1:18.182

Sync Point: 1:18.182

End: 4:31.579

Duration: 3:13.397

☐ Use Subframes

Original Time Stamp: 1:17.563 ▲

User Time Stamp: 1:17.563 ▲

Cancel OK

(a)

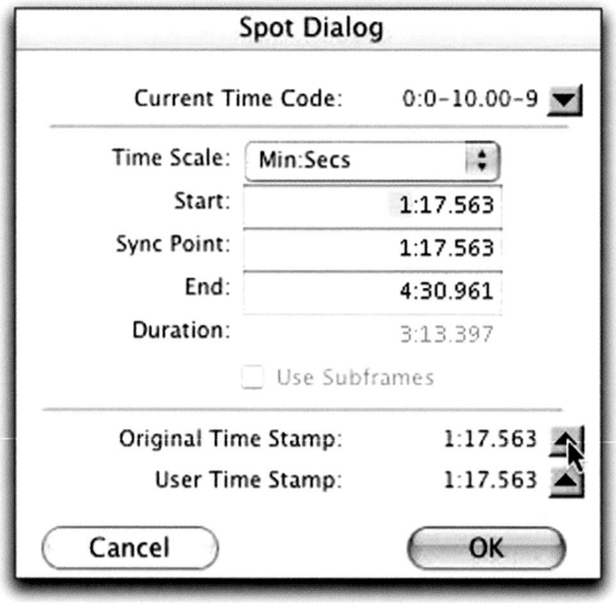

Spot Dialog

Current Time Code: 0:0-10.00-9 ▼

Time Scale: Min:Secs ⬍

Start: 1:17.563

Sync Point: 1:17.563

End: 4:30.961

Duration: 3:13.397

☐ Use Subframes

Original Time Stamp: 1:17.563 ▲

User Time Stamp: 1:17.563 ▲

Cancel OK

(b)

Figure 3.79 The Spot Dialog can take you back where you came from.

Hitting OK on the Spot Dialog will return your audio regions back to where they came from. Whewww.

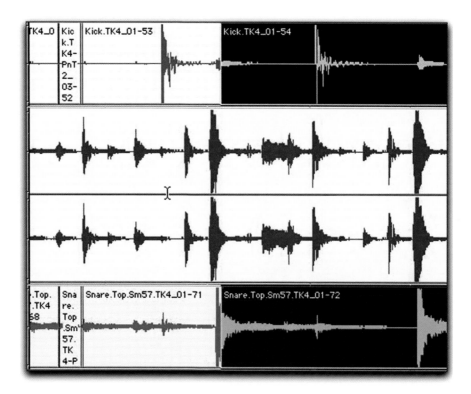

Figure 3.80 The spotted audio is back where it belongs.

Tab to Transients: The good and bad

A few closing thoughts on editing drums. The Tab to Transients feature is often a big time saver when it comes to jumping from one edit or note to the next. And while this is the way you may want to get around a lot of the time, I would also recommend zooming out periodically and scrolling through with your horizontal scroll bar to make sure it hasn't missed a transient entirely.

How much 'feel' to leave in the track

Don't forget that you can be as tight or as loose as you want to be when it comes to pocketing these drums. Remember, we recorded these tracks to a loop, rather than a click, that already has a little bit of a feel that we liked in the first place. As a result, we've been editing this track just a bit tighter than we might have if we were pocketing to a strict metronomic click track.

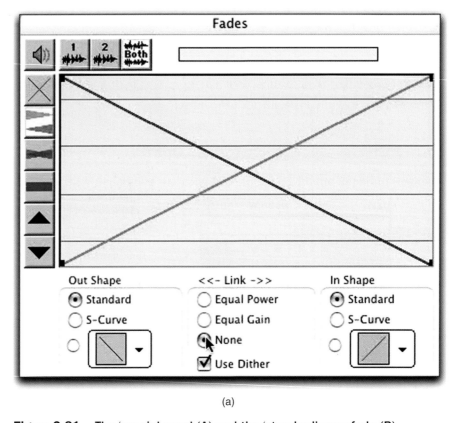

(a)

Figure 3.81 The 'special case' (A) and the 'standard' crossfade (B).

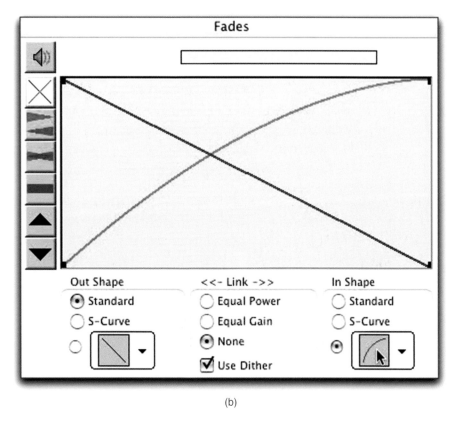

(b)

Figure 3.81   (Continued).

If you need to, refer back to Chapter 2 for the standard crossfade shape that we use for all but the edits between the time-stretched and the non-time-stretched. For those edits, we just go with a standard, no link, linear crossfade. Experience has taught me that this is the best shape for that particular situation.

Go ahead and work your way through the remainder of the drum tracks, and be sure to follow along in your own session with the accompanying DVD for some more great visual examples of pocketing drums!

Wrapping up the drums

After finishing up with your drum pocket, walk away from it for a bit before you sit down to critically listen to it. After spending upwards of a couple hours microediting a song like this, you can easily lose perspective about whether you're improving or destroying it. As you continue to hone your editing skills you will find that you are able to maintain your sense of the song with greater ease and for longer periods of time, but initially it can really be overwhelming. Don't stress about it.

Now that you've had a little while to be away, come back, make sure you're just soloing the drums and the loop/click, and really critically listen through the track several times. This is a valuable skill that is quickly being lost as people are coming out of recording schools by the thousands every year. The ability to zoom in on, cut, and move a waveform around can be learned by monkeys in a zoo, but the ability to make essential judgment calls about the quality of a performance or the musicality of your editing choices is still, at this point, uniquely human. Make sure your track still has that humanity, groove, and the all important vibe.

Once you've got the track cut exactly right, and everything is working well, it's time to lock down our drum track so we don't accidentally make any changes later on.

Since we now know we're not going to be adding any overdubs to the song, we can go ahead and delete those muted regions at the top to the track. They won't be needed any more, so select and delete them.

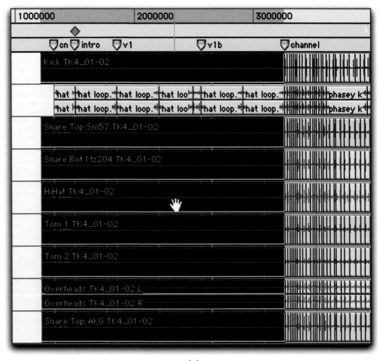

(a)

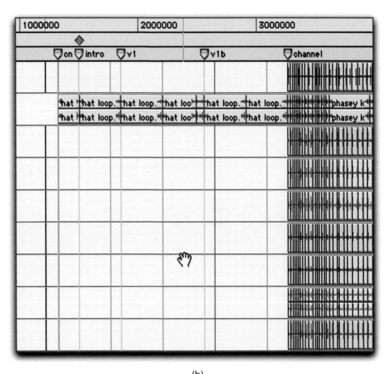

(b)

Figure 3.82 Goodbye to the silence.

The second step is to consolidate your super edited drum tracks into whole files made up of a single region per track. To do this, select all of your drum tracks from the first downbeat to the end of the song. You can obviously accomplish this in a variety of ways, but the easiest is to just triple-click on any of the million drum regions with the Selector tool.

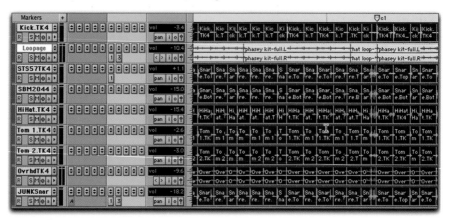

Figure 3.83 Triple-click with the Selector to select all of the grouped drum tracks.

Now choose Consolidate under the newly revised Edit menu to render all of your edited tracks down to individual drum regions.

(a)

Figure 3.84 Consolidate the selected tracks to render them down to individual regions. Option~Shift~3 (Mac)/Alt~Shift~3 (PC).

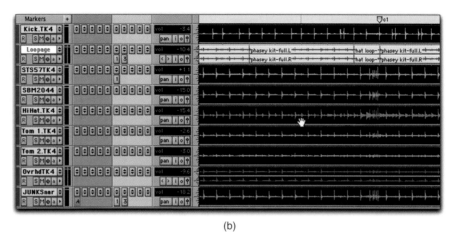

(b)

Figure 3.84 (Continued).

The last two steps to locking our drums both literally and figuratively are to put them on a new playlist with an _M in the name – denoting that this is a Master, completed take, and to use the Pro Tools Lock/Unlock command to lock the regions from being edited.

First, select the drum tracks so they are all highlighted, as in Figure 3.84B. Copy them to the clipboard with Cmd-C (Mac) or Ctrl-C (PC). Next, go to the New playlist option under the Playlist Selector button next to any of the track names.

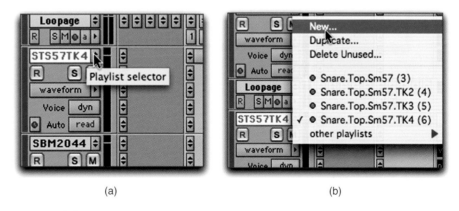

(a)   (b)

Figure 3.85 Creating a new playlist for your Master drum track.

Use your Paste command to paste all of those Master drum parts directly into their tracks, then rename every track and region to track_M (i.e. Kick_M, Snare_M, etc.).

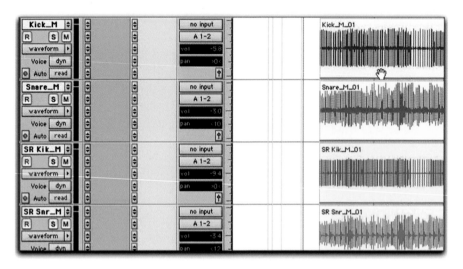

Figure 3.86　All tracks renamed with an _M to denote Master status.

Finally, select the tracks one last time and choose the Lock Regions command by hitting Cmd~L (Mac)/Ctrl~L (PC). A small padlock will appear on every region, allowing you to hear it in the track, but not edit it without first unlocking the region.

Finally, go ahead and do a Save As command and create a new session with the name Home_PT7 Pckt Drums. This will let us know that this session has a complete set of pocketed drums inside, so that even if we somehow mess up our drums during a later part of our edit, mix, or whatever, we can always get back to reimporting the locked, Master pocketed drums from this session.

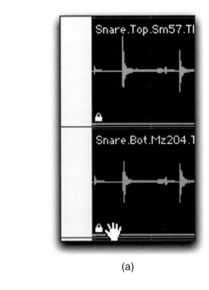

(a)

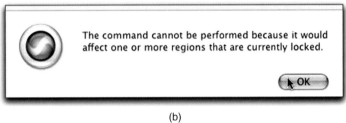

(b)

Figure 3.87 These regions have been locked. Attempting to do anything will result in a Pro Tools no-no message.

Figure 3.88 Saving the pocketed drums with a different session name.

You should end up with a session that looks something like Figure 3.89.

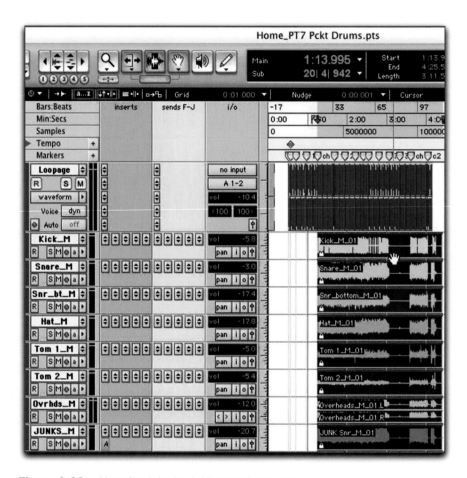

Figure 3.89   Your final, locked, Master drum edit.

Using Beat Detective to save time, money, and headaches

So, our drums are finished and we're ready to move on to our percussion tracks. The typical method for doing percussion overdubs to is save an alternate session and send it off on a hard drive to a professional percussionist, who lays down some amazing and creative tracks. As a result, our first step was to find those sessions and import the percussion tracks into our working edit. This is another spot where you won't need to follow along in your session (until page 135), but take a look with me into the main project folder, under the subfolder 'Old Sessions' for an idea of just how many alternative session files we can generate over the course of a typical song project.

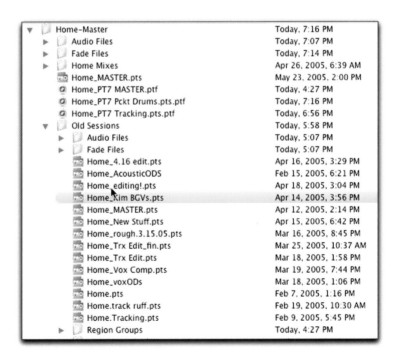

Figure 4.1 Finding our alternative sessions with the percussion tracks.

Importing the percussion tracks

In this particular project, the percussion tracks were done so far back that I don't actually still have the separate percussion project files, but I do have all of the percussion tracks we ended up with in the Home_PT7 Master file.

Like we did earlier, I went under the File menu, Import Session Data, and chose the Home_PT7 Master file to Open.

Looking a bit more closely than we did previously, note that this window shows you the necessary information that we need to find the tracks we're looking for. From the track types of audio, MIDI, instrument, or Master fader, to whether they're mono, stereo, or even surround tracks if we had them, the Source Tracks are the tracks located in the session you're importing from. The Destination column will allow us to add the selected tracks to their own individual 'New Tracks' in the session. If you currently have a track with the exact same name in the new session, Pro Tools HD will allow you to overwrite the existing track in your session with the track from the importee session. Pro Tools LE, unfortunately, will not.

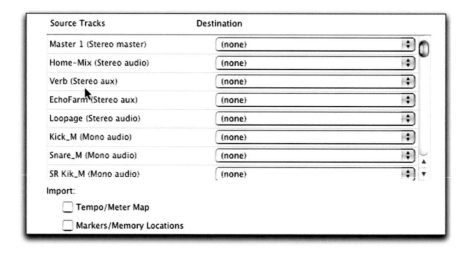

Figure 4.2   The Import Session Data window in Pro Tools LE.

The Audio Media Options

Before we grab our percussion tracks, notice a very important option that we have in this same window under the Audio Media Options. This option tells Pro Tools what to do with any audio files that we're trying to bring from the other session into the current one. Really the primary two options we're ever going to use are the **Link to source media** and **Copy from source media** options. Copy from source media would be the choice if we were importing tracks from another physical hard drive or backup DVD, and we wanted to make sure those tracks were copied to our own Audio Files folder, rather than attempting to play back from the other drive. In this situation, though, since we are just working from one hard drive with all of our audio files and tracks contained within one folder, we will just leave the choice on Link to source media.

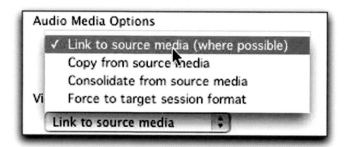

Figure 4.3   Choose **Link to source media** to avoid unnecessary creation of duplicate audio files.

We went down and click/dragged over the tracks from Tamb Hits to Clock 2 and hit OK to import a bunch of percussion tracks to our session.

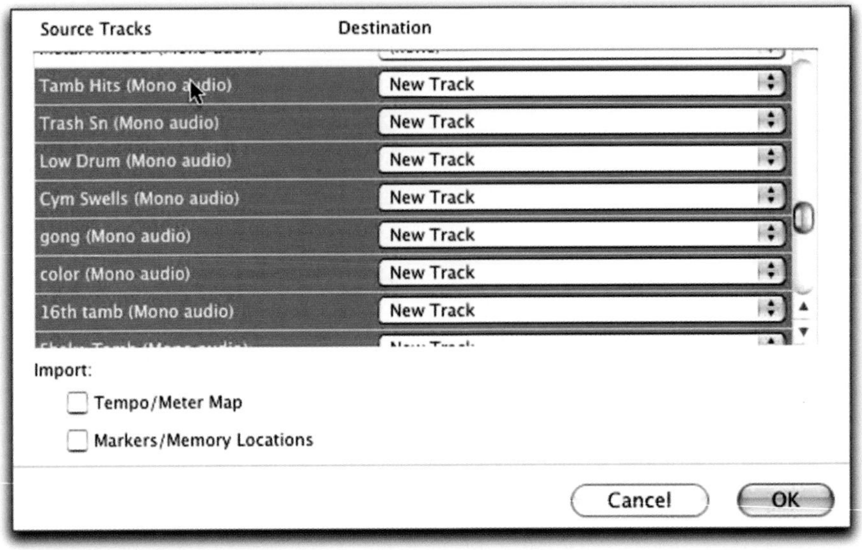

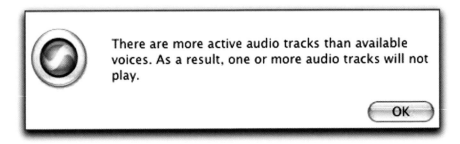

Source Tracks	Destination
Tamb Hits (Mono audio)	New Track
Trash Sn (Mono audio)	New Track
Low Drum (Mono audio)	New Track
Cym Swells (Mono audio)	New Track
gong (Mono audio)	New Track
color (Mono audio)	New Track
16th tamb (Mono audio)	New Track

Import:
☐ Tempo/Meter Map
☐ Markers/Memory Locations

Cancel   OK

Figure 4.4   Importing your percussion tracks.

Unless you're using Pro Tools HD, one will get an error message saying you don't have enough voices to play back this many audio tracks. That's OK. We'll just go over how to pocket one for demonstration purposes, and you can decide how many and which ones you want to use in your final mix of the song.

There are more active audio tracks than available voices. As a result, one or more audio tracks will not play.

OK

Figure 4.5   LE only has 32 audio tracks currently. With the new Music Production toolkit, Digi gives LE users 48 audio tracks. Sweet.

In this case we had far more percussion tracks than we're going to have time to look at, but it would have looked a little something like Figure 4.6. In my opinion, once you've worked with a truly amazing percussionist and seen the level of quality they can add to a song in such a short amount of time, you will simply never go back.

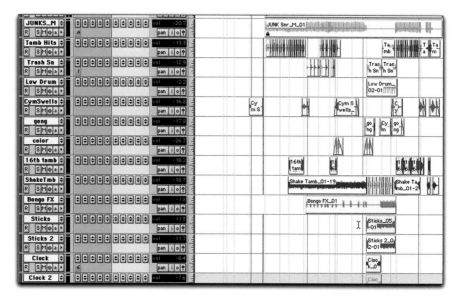

Figure 4.6   Man, I love percussion.

Setting up your percussion Edit window

For the sake of pocketing and Beat Detective, we're only going to take a closer look at that brush track that we made inactive earlier. Since the editing is now a mainstay part of the production process, I think we're going to see if we can't make this brush part really lock into our new, tighter drum groove.

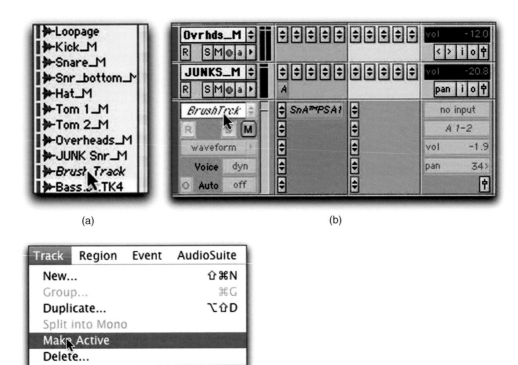

(a)

(b)

(c)

Figure 4.7 Unhiding and reactivating the inactive brush track.

Pocketing the brush track

If you go ahead and play a few seconds of the brush track soloed against the drum tracks, you'll probably notice it's quite a bit off. Obviously, this is because the night he recorded the brush part he was playing to his own groove based on the drums, which, ta dah, we just changed in the previous chapter! Aren't we clever. This has put the brush track a 'slightly-too-far-to-leave-it-alone' 13 milliseconds behind. Not good.

Figure 4.8 Thirteen milliseconds is late enough to feel draggy.

To begin rectifying this, drag the brush track up between the kick and snare tracks, and we'll see what we can do with a little tool called Beat Detective.

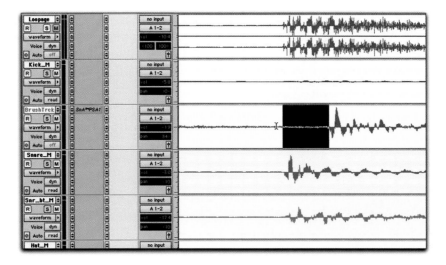

Figure 4.9 The brush track is now quite off from the new pocket of the song. As the Governator Ahnold once said, 'No problemo'.

Beat Detective

Beat Detective is a helpful little application inside of Pro Tools that works well for a variety of types of groove-based instrument editing. You can think of it as basically a MIDI quantizer, but for audio. What it essentially does is analyze the transients of the audio you feed into it, compare that to the beat resolution that you are trying to line it up to, chop the region into what it believes are appropriately sized pieces (with your input, of course), and then decide if each region needs to move forwards or backwards in time to fit the groove you're giving it. About 90% of the time, it makes this decision pretty well. The other 10% or so – well, not so much. If the down-beat of your selection accidentally gets pocketed an eighth note late, the rest of your track gets pocketed against that late note as well. At least until it gets halfway through the track and also pockets another beat an eighth note off – now all of the beats after that are a quarter note off. By the end of your track it gets to a beat or two off, and you then have to go back through the entire track trying to figure out where it got off track, and then set things straight – what a pain.

The main problem with it is that it doesn't really work all that well unless you have very cleanly recorded tracks with well-defined transients. Even then, you tend to spend so much time cleaning up all of the missed transients and bad edit choices it makes that it is often more time-consuming than simply pocketing the song by hand, which produces more musical results anyway (in my humble opinion). In addition, its time reference is based on the Pro Tools tempo, meaning you can't just use it with a track recorded to, say, a drummer's simple metronome recorded into Pro Tools (unless his metronome was being driven by Pro Tools MIDI Time Code). Simply using Beat Detective's default detection to correct the timing on a drum track is like using Auto mode in AutoTune to tune a lead vocal. It may help the subtle problems, but will cause some parts to be worse off than they started. Beyond this, since I'm planning on looking at every note anyway, it tends to only be a time saver for small sections or single tracks.

On TDM and HD systems (and the newly released Music Production Toolkit upgrade for Pro Tools LE), Beat Detective is a multitrack groove editing tool that will allow you to analyze, chop up, apply a groove template, and clean up **multiple** tracks of audio simultaneously, like a drum kit, for example. Currently on a stock Pro Tools LE system you can only utilize Beat Detective on a **single** track's worth of audio at a time.

While some may view this as a disadvantage, I find it really almost works better on the whole for tweaking single tracks like this mono percussive brush track, and can really be a valuable time saver as a result.

Setting up for a good beat detection

To start off, switch over to Grid mode and change your counter method to Bars:Beats. Also, set your grid resolution to quarter notes.

(a)   (b)

Figure 4.10   Grid mode and Bars:Beats will set us up for using Beat Detective.

Figure 4.11 Grid resolution is at quarter notes.

Now zoom out horizontally, and click/drag over just the first region of brush track that starts around bar 34. This brush track has regions that play at different times over the course of the song, but to improve Beat Detective's abilities, we're only going to select it one region at a time. This is really helpful for any application of Beat Detective. For whatever reason, giving it a smaller chunk of time to deal with helps it improve its detection accuracy. In this case we're starting by trying to grab an entire chorus – which may actually be too much. But hey, let's live dangerously for a minute, whaddaya say?

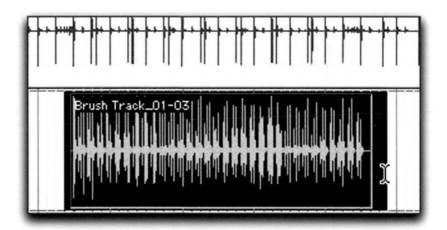

Figure 4.12 Selecting only a piece of audio at a time helps improve Beat Detective's ability to detect – beats. Ahem.

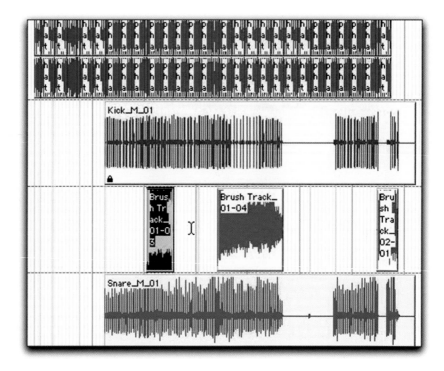

Figure 4.13 Let's just start with a chorus and see how the night goes, shall we?

Launching and using Beat Detective to pocket the brush

Now go to Beat Detective's new Pro Tools 7 location under the Edit window and choose Beat Detective, or just use Cmd~8 (Mac)/Ctrl~8 (PC).

Figure 4.14 Opening Beat Detective under its new menu location.

This will launch the Beat Detective window in Pro Tools. Since we already have our first audio selection made, we need to tell Beat Detective that this is the piece we're going to be working with, so we click on Capture Selection.

Figure 4.15   Click Capture Selection to load our brush region into Beat Detective.

Make sure it has the proper Start and End Bars, the time signature you're using for this song, and the resolution of the beats contained in this particular track, which in our case is 1/16th notes. This should follow whatever time signature you've used for the click, but you should make sure.

Figure 4.16   Making sure Beat Detective has the right idea.

Now switch to the Bar | Beat Marker Generation button and you should see a bunch of marker lines filling your audio selection. Drag your Sensitivity slider all the way to 0% (left) to make them all go away.

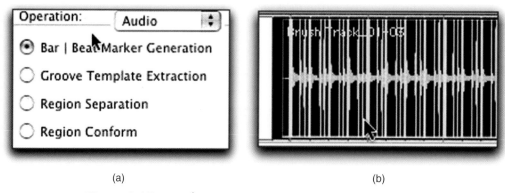

(a)   (b)

Figure 4.17   Bar | Beat Marker Generation divides your region up into its best guess of your notes. Let's help it out a bit more.

Switch your Analysis option to High Emphasis and click the Analyze button to have Beat Detective analyze your track.

Figure 4.18   Getting Beat Detective ready to 'do its thing'.

Once the audio is analyzed your Sensitivity bar will become active again, and you need to slowly drag it to the right until you see the marker bars starting to mark out the transients of your audio selection. You want to turn the Sensitivity up until it marks out every one, or nearly every one, of the audio transients.

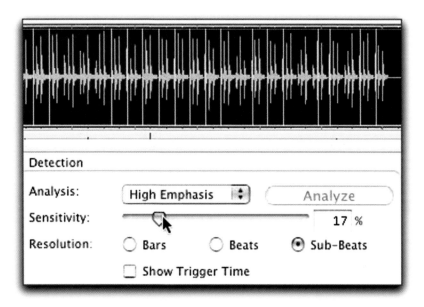

Figure 4.19 Low sensitivity misses a lot of the markers. Keep turning it up until it gets the majority of them.

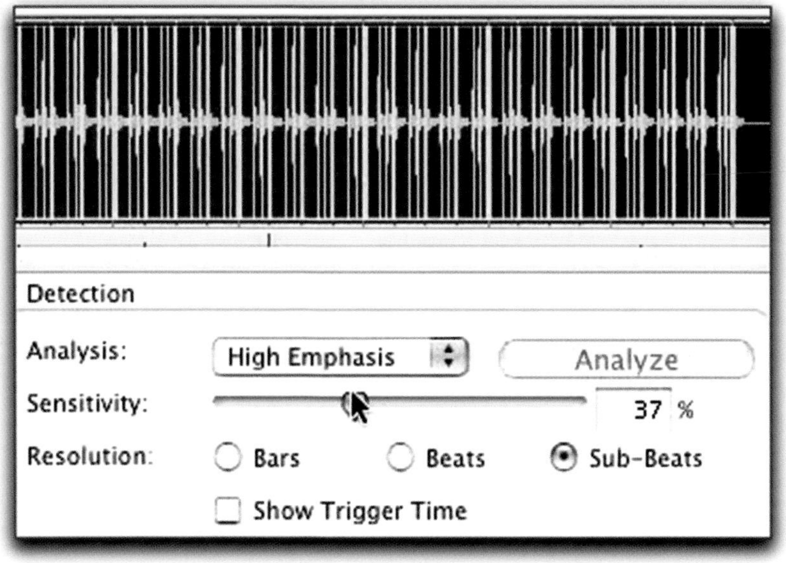

Figure 4.20 The right Sensitivity level looks to be around 37%.

Once your Sensitivity hits around 37%, it looks like we should have virtually all of our transients marked. You certainly don't want to go any higher or you will start to get double triggers on the transients. Double triggers happen when the Sensitivity is set too high and Beat Detective tries to cut one note into two or three trigger pieces, and you don't want that to happen. Just to be sure, zoom in a bit closer to make sure all of the transients have one, and only one, trigger marked.

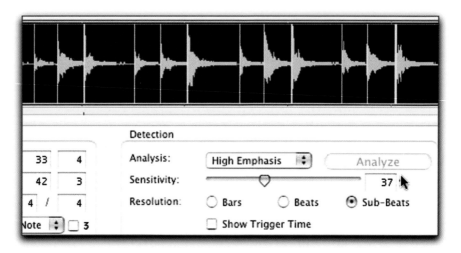

Figure 4.21   Double-checking on Beat Detective. This looks good.

Switch from the Bar | Beat Marker button to the Region Separation button. Note the Separate button that appears in the lower left corner of the window. Click Separate to have Beat Detective create edits at all of the previously created trigger points. Zooming back in now will show you that it has inserted edits at all of the trigger points, creating a plethora of new regions, each with their own unique region name.

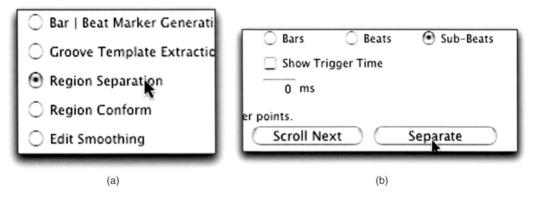

(a)   (b)

Figure 4.22   Region Separation and Separate commands.

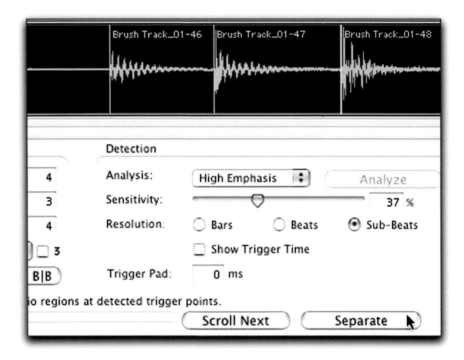

Figure 4.23   Separating the region into tiny pieces of audio.

Now we move on to the Region Conform button. If you have done any kind of MIDI quantizing at all, you should feel immediately comfortable with the options that now show at the right of the Beat Detective window. You can choose your conforming strength, make it exclude regions within a given percentage, or even add a little swing feel to your brush groove – not exactly what we're looking for.

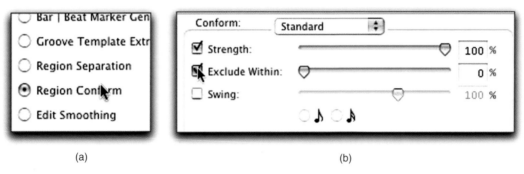

(a)   (b)

Figure 4.24   Just like MIDI, Beat Detective is ready to quantize.

Honestly, the first thing I do is to leave every option (Strength, Exclude Within, and Swing) turned 'off', and click Conform, just to see what happens. When you click Conform, you should see the selected regions move just a bit.

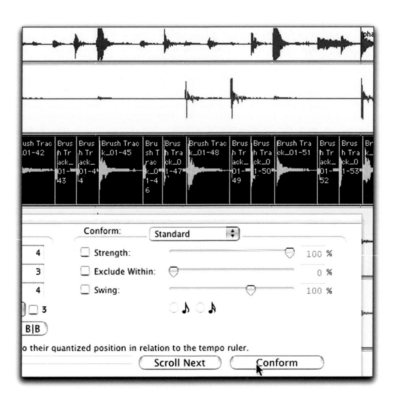

Figure 4.25   Before conforming the regions.

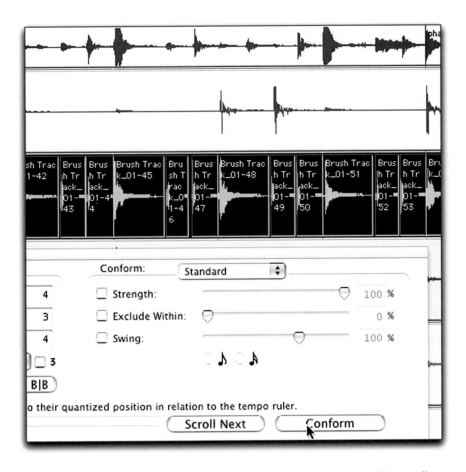

Figure 4.26 After conforming the regions. Notice they slid a little earlier in time.

At this point there is nothing to do but hit Play and see how the Beat Detected brush track fits in with the drums.

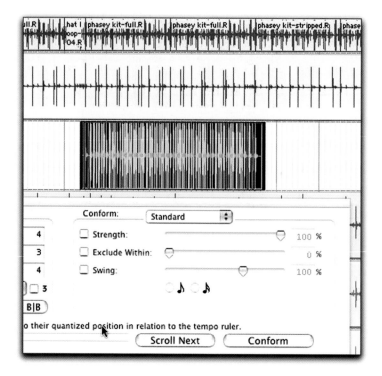

Figure 4.27   Zooming out and listening in. Did we improve things?

Keeping your frame of reference: Did you improve the track?

If you need a frame of reference, hit your Undo command to quickly toggle back to the pre-beat detected brush track for comparison. Then, just choose Redo under the Edit menu to go back to the new track. I like to toggle between these repeatedly to compare my previous and existing tracks.

Sure enough, the un-Beat Detectived (to create a new word) track starts to get late after a few bars in. In general, I like what it's doing to the track, so I'm going to make sure I redo to where I have the new edit, and zoom in to take a closer look at what Beat Detective has done to my track.

Edit smoothing and filling gaps: The right choice

Notice the huge gaps, holes, and complete lack of crossfades between nearly all of my slices of audio. If we were to solo this track, those would definitely be audible and add a number of clicks and pops to my tracks.

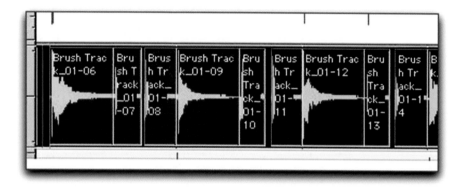

Figure 4.28   Notice the huge gaps between some slices.

In order to take care of this, click over to the Edit Smoothing button and look at our two options: **Fill Gaps** and **Fill And Crossfade**.

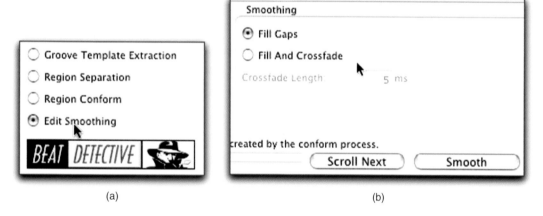

(a)   (b)

Figure 4.29   Edit Smoothing for filling in the holes left by Beat Detective.

If we just choose to fill in the gaps created by Beat Detective, Pro Tools just automatically trims the audio back from the second region to fill in the gaps created.

149

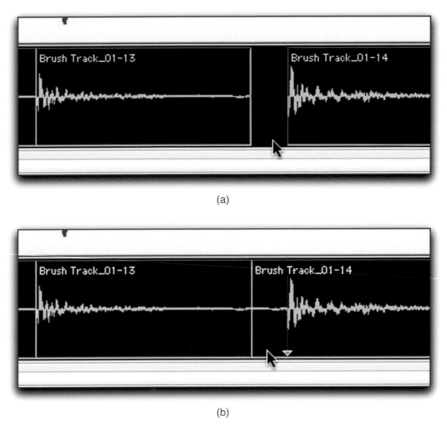

(a)

(b)

Figure 4.30 The result of the Fill Gaps command.

Obviously, the Fill And Crossfade option will do something similar, except it allows us to determine the length of the crossfade that it will automatically put at each edit point it creates.

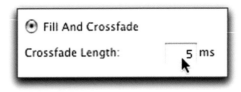

Figure 4.31 A 5-millisecond crossfade is a good starting point.

Choose Fill And Crossfade, click Smooth, then solo and listen back to *just* the brush track to critically listen to the edits. Use the same method of Undo/Redo we just used for the Fill Gaps command to make sure you're improving upon the original performance.

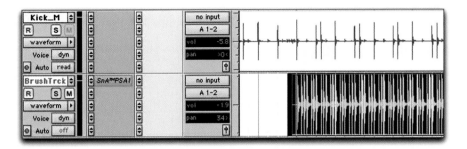

Figure 4.32 Just checking the brush track to make sure those crossfades helped.

Fitting it in the mix: Keeping your time priorities straight

It sounds pretty good to me, which brings me to the idea of perspective when you're doing a lot of pocketing and beat detecting. This brush track is at best going to be a secondary percussion element that is only a small part of a greater drum track. It's not a piano track, vocal track, or anything that's going to be really prominent in the mix. What I'm mainly concerned about in the interest of time, quality, and everything, is that the more exposed the instrument is, the more time and energy I'm going to spend on it. The less exposed an instrument is – for example, a hip, effected brush snare that's going to be tucked back in the mix 10–12 dB – the less I'm going to worry about it. For a track like that, all I'm going to be concerned about is the timing, and making sure there are no clicks and pops to be heard throughout the track.

And so Beat Detective has saved us from ourselves in this small situation. For example, if I was going to go through and pocket this track manually, note by note for the entire chorus, it probably could have taken 30–45 minutes. With Beat Detective we were able to do it in 5 minutes and save ourselves a pretty substantial amount of time. If this were a brush track that ran intro to outro throughout the entire song, Beat Detective would allow us to pocket the track in 10–12 minutes, rather than the couple of hours it could take to do it by hand. Very nice indeed.

To really put it all in context, go ahead and solo your loop back in along with the drums and give it a play through with your brush track volume turned up just a bit for posterity and critical listening purposes.

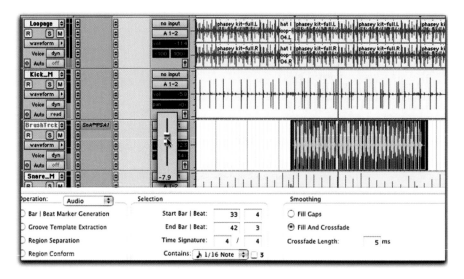

Figure 4.33 Listening to our brush handiwork with the drums and loop.

Special cases in Beat Detective and how to address them

Fantastic! Now that that's done and our confidence in Beat Detective with this source material is up, let's jump down to the next section and repeat the process while looking for any special cases that might leap out and bite us.

First, select over the entire chunk of audio starting around bar 56 to bar 77, go to your Bar | Beat Marker Generation tab and choose Capture Selection to load the audio into Beat Detective's engine.

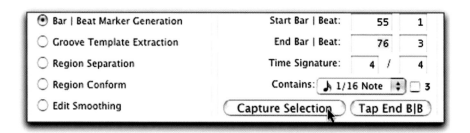

Figure 4.34 Capturing our selection.

Select High Emphasis, click Analyze, then drag your Sensitivity slider down to zero, then back up until you see your trigger points dropping in front of all of the wave transients.

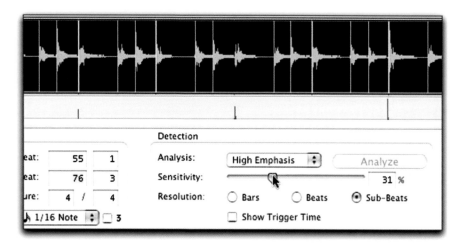

Figure 4.35 Analyzing and setting the Sensitivity level.

At a close zoom level, click on your scroll bar to scroll through the track to make sure all of the transients are triggered, and don't have more than one trigger apiece.

Like before, go to the Region Separation button, and let's look at one of the preferences that we didn't use last time, but may need in the future. When zoomed in tight on a waveform and trigger, notice that Beat Detective puts its trigger right at the front of what it perceives as the transient, right at the front of the attack, when the Trigger Pad option is set to 0 ms.

The Trigger Pad option

Now, if we set our Trigger Pad option to say, 10 ms, and choose Separate, note that it puts our actual edit 10 ms before what it perceives to be the actual trigger time.

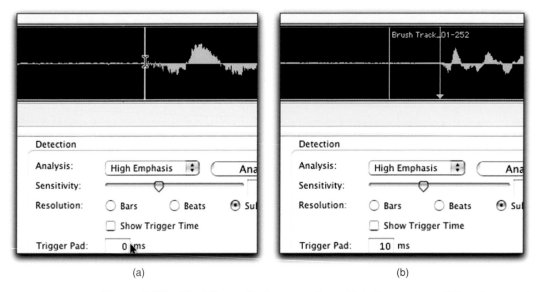

Figure 4.36 The Trigger Pad causes the edit to be made well in advance of the perceived trigger point.

The reason we will often utilize a Trigger Pad when using Beat Detective is because it leaves a little bit of an audio pad, so that when Beat Detective actually moves all of the notes around, it doesn't accidentally overlap any beats to create a double transient. What can often happen is that when it moves notes later in time, and then goes back with a Fill Gaps or Fill And Crossfade, similar to what we saw when editing the drums, it can overlap two of the same fronts of a beat in the middle of a crossfade, resulting in an audible bump in the sound. Turning on 5–10 ms of Trigger Pad helps prevent this. Give the chapter on drums another read for more on this topic.

Now, this second time we've selected a much bigger chunk of material than we chose during our first Beat Detectivizing (there I go with another new word again). So let's see what kinds of problems we encounter when we just quickly run through the process. Hit the Separate button, switch to the Region Conform tab, and choose Conform.

(a)   (b)   (c)

Figure 4.37 Separate, Region Conform, and Conform – let's see how good Beat Detective *really* is when left to its own devices.

Ouch. If you're like me, you probably just got this warning message that should automatically make you wonder exactly what is going on here. Let's humor it and choose Conform anyway.

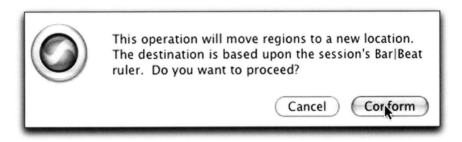

Figure 4.38 Like Hal from *2001: A Space Odyssey*, Beat Detective is letting us know it thinks it knows best.

Sure enough, Beat Detective moved our entire set of selected regions earlier by several beats. Ummm. That's not good. Go ahead and undo.

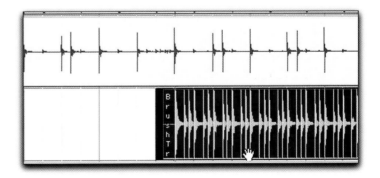

Figure 4.39 The old location.

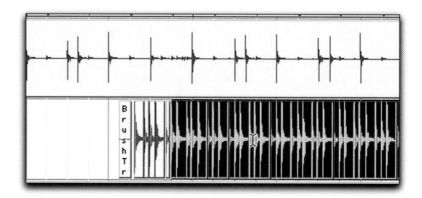

Figure 4.40 Where Beat Detective moved our regions to – no good.

Once you undo, take a peek over at the Start Bar and End Bar options. If you hit the Capture Selection button now, you will see that somehow we got our start and end times screwed up in Beat Detective, because we neglected to capture our selection at the appropriate time. This is just a little heads up for any sort of warning message that may come at you in Pro Tools. Always be sure to read them to see what they're telling you, or you may find yourself up a creek without a paddle. When it comes to Beat Detective and your selected audio, it can seriously throw things out of whack if either of your start or end times is off.

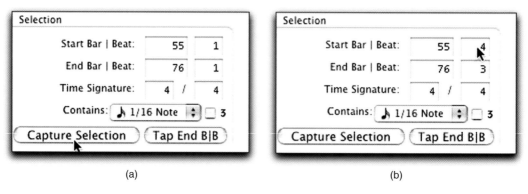

Figure 4.41 Make sure you choose Capture Selection first. Otherwise, Beat Detective may be operating under false assumptions, which isn't good for a detective of any kind.

What you should take away from this error is that while Beat Detective can be a great time saver, as we have already seen, it can also leave you wondering – why the heck did it do that? My rule of thumb is that if I choose

Conform and see Beat Detective move the audio any more than just a minor, minor shift, something has likely gone wrong.

A quick Beat Detective recap

So, to catch back up to where we should be: (1) Undo to the unedited audio, (2) choose Capture Selection, (3) hit the Analyze button, (4) adjust your Sensitivity and your Analysis to High Emphasis, (5) switch to Region Separation, (6) hit Separate, (7) switch to Region Conform, (8) hit Conform. Let's see if that took care of it.

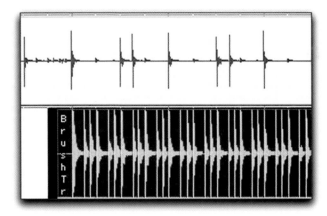

Figure 4.42   Before conforming the region.

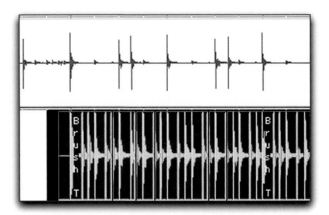

Figure 4.43   After conforming the region. Notice there is no dramatic shift of the selected audio in time.

Indeed, it looks a lot better than the previous attempt. Since we're just being so adventurous and trusting of the robotic rhythmic brain of Beat Detective, let's go wild for a minute and move on to our Edit Smoothing operation.

Figure 4.44 Time to smooth out those edits.

Still being adventurous, choose the Fill And Crossfade command and set a Crossfade Length of 5 milliseconds. Boy howdy, are we walking on the wild side or what?

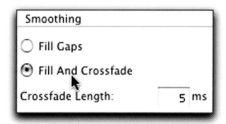

Figure 4.45 Fill And Crossfade.

Hitting Smooth will give you a progress bar while we anxiously await to see whether it works or not.

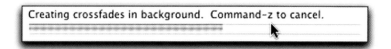

Figure 4.46 Beat Detective's working. Do we dare to hope?

And now it's done. At this point, we could either move forward assuming that Beat Detective has performed a proper edit and perfectly aligned everything to an even grid, or not. Personally, I trust my ears and your ears over the programmed ears of an oft flakey program – so I say go ahead and give it a listen to make sure it feels good and is in the proper time. To do this, I'd solo and turn up the brush track louder than it will be in the track, say to −1 dB. Also solo the drums and make sure the loop track is muted. Turn on a pre-roll setting of around two bars and let's listen to the track.

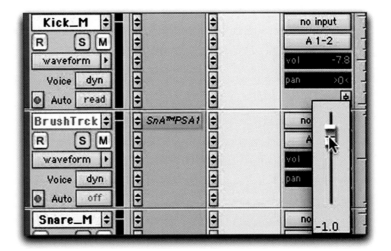

Figure 4.47 Turning up the brush track to check up on Beat Detective's work.

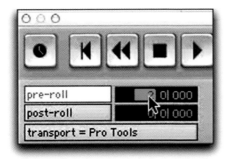

Figure 4.48 Setting a two-bar pre-roll.

Fixing clashes between Beat Detective and the master drum take

Around bar 57, beats 2 and 3, the drummer gets a little funky on the snare and it initially sounds like it may clash with our brush track. If you want to unsolo the brush, you can hear it's just the drummer being creative. The question at this point becomes whether we're going to adjust the brush track or the drums. Since we've basically already signed off on the drums as a master track, we're likely going to just fix the brush track. While we're noting problems, zoom in close and notice that the brush track also appears to be quite a bit on top of our established beat.

Make sure you're in Grid mode and take a look at the Bars:Beats ruler at the top of the Edit window. This Pro Tools session was recorded at the

159

(ironically) default tempo of 120 bpm. But you'll notice our click/loop track is slightly behind the downbeat of the Pro Tools grid marker for beat 1. Check out Figure 4.49 for a visual.

The problem we're encountering is that both our loop and hence our already pocketed drums are just slightly behind the Pro Tools internal click/grid, but Beat Detective is pocketing everything *exactly* on the Pro Tools internal click/grid – hence making the brush track now just a bit too on top of our beat.

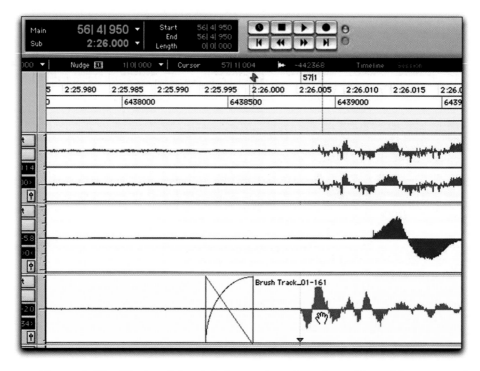

Figure 4.49   The brush track is too on top of the beat. It's making our track feel rushed.

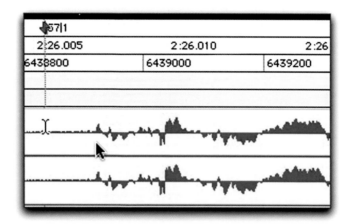

Figure 4.50   Aha, here's why. The insertion 'I' shows the Pro Tools grid downbeat at 57 | 1, while the arrow is pointing out that our click/loop track that we pocketed our drums to is slightly behind the internal Pro Tools grid.

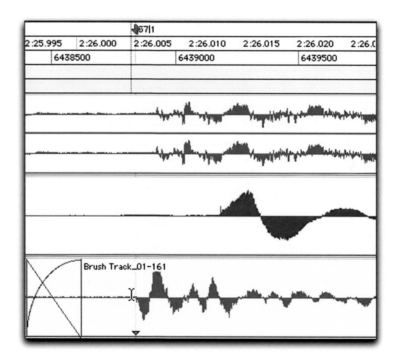

Figure 4.51   Beat Detective doesn't take into account that the rest of our track is slightly behind the beat.

Now switch your Nudge value from Bars:Beats to Min:Secs: 1 millisecond, and we're going to micro-nudge all of these beats to line up with our pocketed drums and loop.

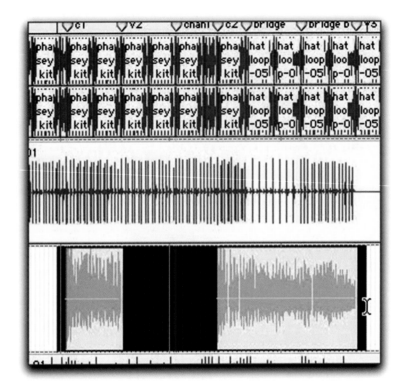

Figure 4.52　Selecting all of the brush track regions.

The solution? When you're satisfied that the track is otherwise in the pocket, zoom out to where you can see all of the brush track regions, and click/drag to select over them all.

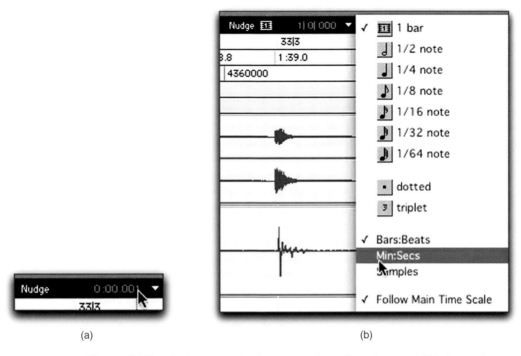

(a)   (b)

Figure 4.53   Switch your Nudge menu from Bars:Beats to Min:Secs, then choose it again to select 1 millisecond.

Now, using the + and − keys on your keyboard's numeric keypad, zoom in close and nudge that entire track until your brush track transients find their way slightly behind the drum transients. This should end up being a few milliseconds worth of nudging to get your brush track looking like Figure 4.54. All of your percussion elements should be made to find themselves slightly behind your drums.

If they were pocketed right by Beat Detective and you made sure they fit the track already, this should just be a matter of nudging them all at once from being on top of the beat to being slid in just behind the drums for a sweet, tasty, laid-back vibe.

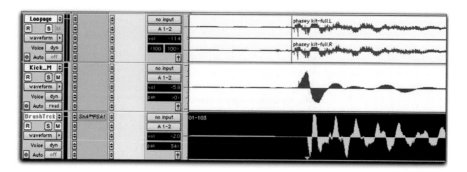

Figure 4.54   Percussion elements should be pocketed just behind the drums to avoid rushing the song.

Now go back and listen to the part again. Notice that just nudging our brush track in behind the drums rather than on top of them has both made the track feel a *lot* better, and it has minimized that timing clash with the snare at bar 57. While it may not have completely erased the conflict with our drummer's creative snare part, it has made it so you'd almost never notice it in the track. However, since this is Multi Platinum Pro Tools, rather than, I don't know, Multi hundreds of copies Pro Tools, let's fix it by just nudging the one offending brush hit.

Zoom in on the brush track at bar 57, beat 3 and you'll see just the one offending brush hit is still ahead of the associated snare beat. Just select the one brush hit and use your + and − keys to nudge it into place.

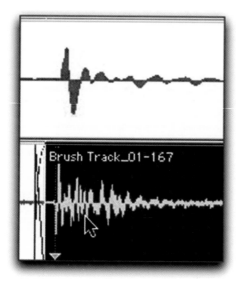

Figure 4.55   This one brush is still on top.

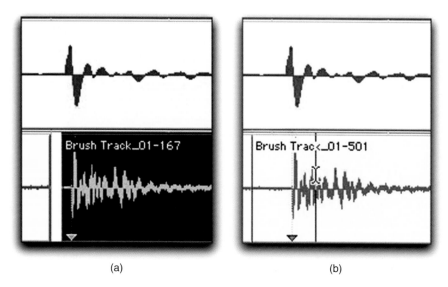

(a)   (b)

Figure 4.56   Nudge it behind the drum beat and trim the edit point back to fill in the gap. Drop in a crossfade and move on to the next note.

Once it's in the pocket behind the kick drum, trim your edit point back and make a quick crossfade just like we did during the drum pocket. That cleans that one up!

Of course, once you edit one, you can't just stop there. Notice that the next hit also needs pocketing behind the drums. No problem. Just select the fade at the end of it and delete the fade. Then, select this brush hit itself and nudge it into place. Trim the front and back to where they need to be and crossfade in and out of it. Simple as that.

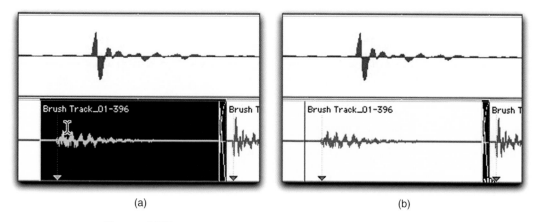

(a)   (b)

Figure 4.57   Selecting the note and deleting its fade out.

165

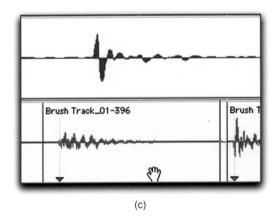

(c)

Figure 4.57 (Continued).

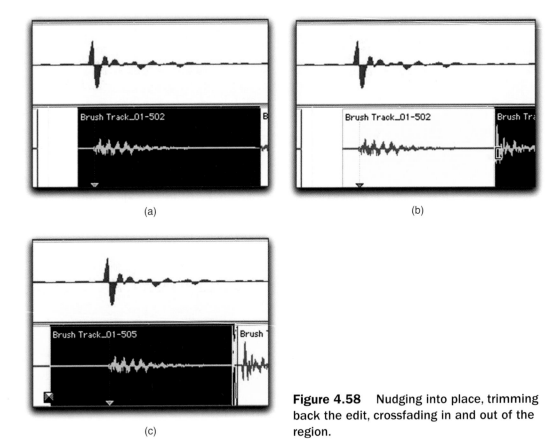

(a)

(b)

(c)

Figure 4.58 Nudging into place, trimming back the edit, crossfading in and out of the region.

Basically, as long as Beat Detective puts me in the ballpark, it's a lot easier to go through and nudge the occasional note that needs tweaking than it is to manually go through the track note by note if we don't have to. After you've completed the above edits, you should no longer have any clash between the snare and brush track on bar 57.

Using batch fades with Beat Detective

Before we move on to the bass guitar, let's look at my approach to batch fades.

If I wish to have more control over the types of fades used over a broad cross-section of edits (like when we have just beat detected a large brush track region, for example), rather than Edit Smoothing with the Fill And Crossfade command, I can just use the Fill Gaps command and then execute a Batch Fade command across the whole selection. Here's my approach.

Say we switch Pro Tools back to Grid mode, select the next brush track region from about 104|4|000 to 111|3|000, and center the selection on the screen with Option~F (Mac)/Alt~F (PC).

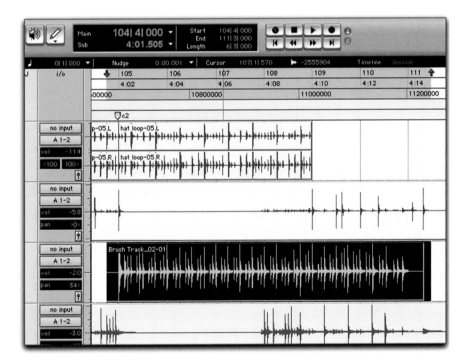

Figure 4.59 Setting up for a quick Beat Detective and batch crossfade with Option-F.

Now we open up Beat Detective, Capture Selection, Analyze, Region Separation, Separate, Region Conform, Conform, Edit Smoothing, Fill Gaps, Smooth – leaving us with a million little regions that we want to put another million fades across, without doing them one at a time.

This is where our Batch Fades command comes in handy, and it's as simple as hitting Cmd-F (Mac)/Ctrl-F (PC) to bring up a special Batch Fades dialog box, ready to grant us three wishes: Pre-splice, Centered, or Post-splice.

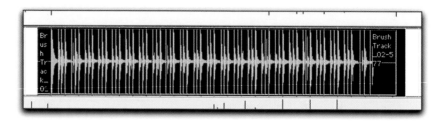

Figure 4.60 Our third beat detected brush region, waiting for a Batch Fade command.

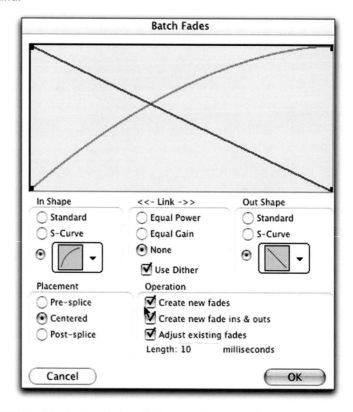

Figure 4.61 The Batch Fades dialog box. Note the Placement and Operation options.

Obviously in this window we can create the standard crossfade shape that we discussed earlier for the majority of our edits. In addition, under the Operation options we can tell the dialog box to adjust any existing fades that may already be found in our selection of edits, whether to create new fades, etc. I tend to just leave all of these checked. When it comes to the Placement options, I always leave Pre-splice checked and here's a visual example of why.

Using the Pre-splice option to avoid double transients

When you choose Centered, Pro Tools will create the crossfade centered directly around your edit point, just like it does when using the standard Crossfade tool. As we can see if we try this around 107|2, this can lead to a double transient on occasion when the later region has had to be moved to the later in time by a significant amount.

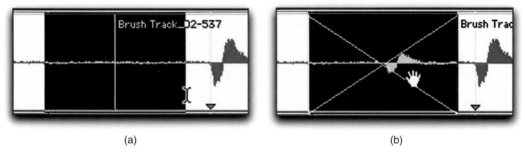

(a)   (b)

Figure 4.62 Centered will center your crossfades around the edit points – sometimes leading to a double transient bump.

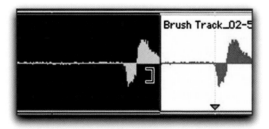

Figure 4.63 Trimming the edit reveals the transient on the right was formerly much earlier – so a centered crossfade won't cut it.

Pre-splice, on the other hand, will set the end of your crossfade to the location of your edit point, rather than the center – thus avoiding a dreaded double transient in the audio, and continuing to save valuable time.

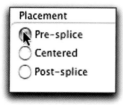

Figure 4.64   Use Pre-splice.

Figure 4.65   Pre-splice ends the crossfade at the edit, rather than centering the crossfade there.

To reiterate, the whole point of using Beat Detective is to save you time. Pocketing an entire brush track like this by hand could take you upwards of an hour and a half. The fact that it is only going to play a minor part in the feel of the song, however, dictates that we spend as little time on it as necessary to get it to the feel we want to get it to, and of course that's where a judicious use of Beat Detective comes into play.

The lazy way out: Avoiding Beat Detective with Copy and Paste

Which brings us to the final method of approaching a track like this, which you can view as the lazy person's method, or the ingenious invention of an editor gone crazy after hours of manually pocketing percussion tracks.

Since our whole Pro Tools track was recorded to a click anyway, to absolutely save the most time and still get the benefits of a low-in-the-mix, pocketed brush track, we could just manually pocket the first two bars of the brush, and then duplicate or repeat it as necessary throughout the song.

This would be done by going back into Grid mode, setting your grid amount to one bar, and selecting the first two bars of the manually pocketed brush track. Then, use Option-R (Mac)/Alt-R (PC) to bring up your Repeat box, and set your pocketed two-bar loop to repeat as many times as you want the brush to last. That's it. But then, that would be a cop-out – wouldn't it?).

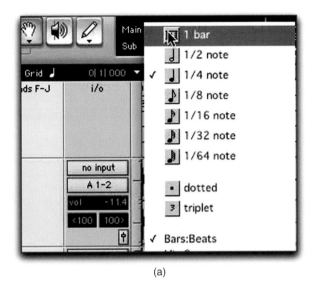

(a)

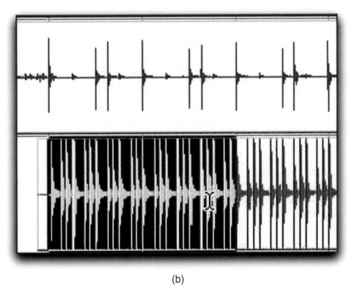

(b)

Figure 4.66 Setting your Grid to 1 bar and selecting two bars of pocketed brush track.

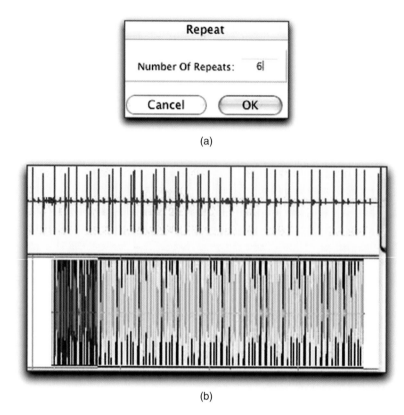

Figure 4.67 Choosing a number of repeats and making the loop last as long as you want.

Ultimately, whenever you have tracked to a click based on the Pro Tools tempo/grid, it opens up a whole avenue of options that you will otherwise be unable to take advantage of. In addition, once your drums are pocketed against this grid, it just becomes easier and easier as we work our way through the instrument chain to create a tighter, more polished perform-ance and that solid feel all the major records are going for. Whether you're working in Rock, Pop, Country or whatever, having your song tracked to a consistent tempo can only lead to a better performance, and faster, more creative, and more abundant editing choices. So now, let's move on to take a look at the bass.

Getting the bass player on time – and not just to rehearsal

Now that the drums are pocketed and the loop has been beat detected, we're going to zoom back in tight and go after the groovy. vibe of the bass guitar.

In addition, since I know a lot of you skipped right over Chapters 1 and 2 to get to the 'good stuff' (you know who you are), let's recap a few common settings. Feel free to speed read through the next few pages if you're a good student who started at page one.

Bass guitar pocketing: The setup

One of the key elements of every great rhythm section is the ability of the bass guitar and drummer to wed their sounds together, building a rock-solid, unyielding, rhythmic force. While truly great players can lock into a stunning rhythm sound time and time again, engineers have always used little tricks to further gel the 'thump' of the bass guitar to the 'thud' of the kick drum. Historically, this has been done by keying an expander on the bass guitar channel to an output from the kick drum track. This way, when the kick drum would 'kick' the expander would open up, allowing the meat of the bass to blast through, sounding locked in and tight with the transient of the drum.

Now, thanks to the miracle of nonlinear editing, we're going to expand on this age-old concept by physically pocketing the bass guitar against the backdrop of the drums and click loop. The result will be a tight, focused low-end that leaps out of your studio monitors and terrorizes your sub (not to mention your neighbor's cat).

In this session, since we pocketed all of the drums to the loop initially, we're now going to pocket the bass guitar to the drums. Specifically, the kick, snare, and loop will provide our pocketing backdrop.

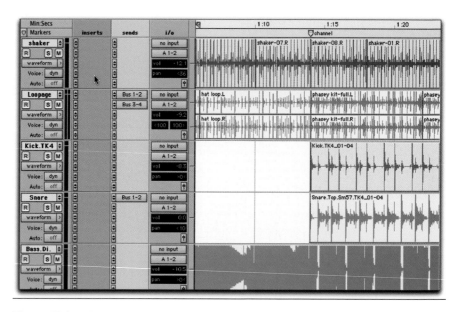

Figure 5.1 Bass will use kick, snare, and loop to pocket against.

To start, let's turn off auto-scrolling.

Figure 5.2 Disabling auto-scrolling.

We disable auto-scrolling because much of our editing will be done while tightly zoomed in near the sample level. As I'm sure you've noticed when editing at a tight zoom before, if auto-scrolling is enabled, the screen will constantly try to follow our timeline every time we hit Stop or Play. By disabling auto-scroll while pocketing, Pro Tools will allow us to stay focused on the soundbite we're nudging about.

Of course, make sure that your session is in Slip, rather than Grid, mode.

Figure 5.3 Engage Slip mode for pocketing.

Slip mode or Grid? Using your ears instead of your eyes

Like we've said before, while Grid mode is often a valuable method of editing for sessions involving a truckload of loops (as in Hip-Hop, Rap, or many forms of electronic music), or even occasionally drums like the previous chapters, for Country, Rock, or any other genre that requires a more 'human' feel, Grid mode will tend to make your song sound a bit more C3P0 than Princess Leia.

> Remembering back to our drum pocketing rule of thumb, we never want to pocket anything dead on top of (i.e. transients exactly matching up with) the transient of the click itself. Since we always pocket our drums first, if you pocket your drums (and, henceforth, all of the rest of your instrumentation) directly on top of the click, you will tend to find you've dehumanized and mechanized it. I've found that it's always good to leave 5–15 milliseconds of flex time that the transients of the drum hits can be behind (later than) the transients of the click or loop that you track to. More than 15 milliseconds and the hit will definitely feel late. Less than five and you've become everything we hate about MIDI. Only if the song required a rushed or anxious feel might you ever consider pocketing sounds directly on top of or even slightly ahead of the beat. Regardless, your editing should always be about fitting the performance to the needs of the song.

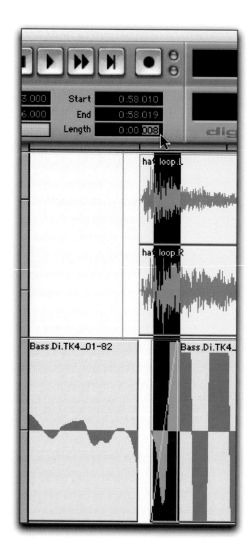

Figure 5.4 Five to eight milliseconds back is a relaxed feel.

The next step is to prep our screen and track layout for the smoothest and most efficient workflow. Doing this now will minimize the amount of clicking and scrolling required during the editing process – and you'll also look like a Pro Tools guru as well (which can't ever be a bad thing, can it?).

The Show/Hide bin: Focusing your edit

Start by Option-clicking (Mac) or Ctrl-clicking (PC) on any of the tracks in your Show/Hide box. Or just click on the Show/Hide bar and choose 'Hide All Tracks'.

Figure 5.5 Show/Hide box to the left of the Edit window.

This will make all of your tracks disappear from Pro Tools.

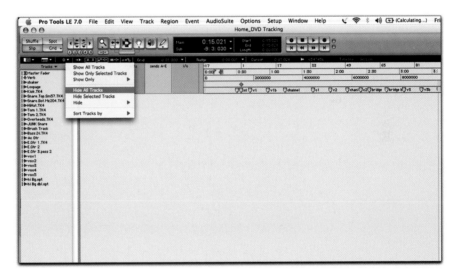

Figure 5.6 All tracks hidden. Only show tracks necessary for pocketing each instrument.

Now, before you run out the door in an angered huff to return this book for deleting your entire recording session, simply click on the kick, snare, and bass guitar tracks to make them reappear. We're not going to clutter our screen with unnecessary bits like guitars, vocals, and kazoo, when we're really just needing to pump up the bass – so to speak.

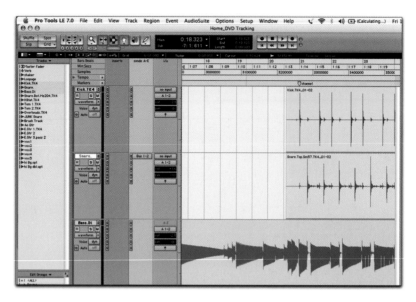

Figure 5.7 Showing kick, snare, and bass is a typical bass pocketing setup.

Now, drag the bass guitar between the kick and snare tracks, zoom in close to the first bass note, and click on the Expand Waveform button to get a much better look at all of the waveform transients.

Figure 5.8 Use Waveform Zoom to magnify waves for easier visual alignment.

Utilizing the Zoom Waveform – grow transients, grow

At first glance, this may look like it is making the audio files louder. In actuality, it's just a visual tool that helps us to find edit points on weakly recorded audio. Even if your track is as hot as a New Orleans tamale, you may still find that a zoomed waveform is a happy waveform, and a happy waveform is more easily pocketed.

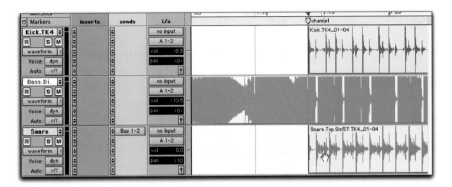

Figure 5.9   Place bass between kick and snare for pocketing to either.

But lo, what's this? At the top of our song you may have noticed that the drums have not yet come in, but the bass guitar (and everyone else) is jiving along with the loop track for a tres cool, vibey intro.

This is a common production technique that is currently used in a lot of intros, bridges, and breakdowns. Some might say that it's less critical to pocket the instrumentation during these more 'swimmy' sections of the song. In my opinion, pocketing is what makes many of these sections work so well, even in the absence of a real drum kit. As a result, go ahead and 'Show' your loop track by clicking it in the Show/Hide window. Also, temporarily move the kick and snare tracks below the bass to get them out of the way.

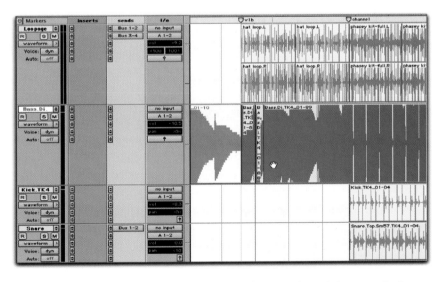

Figure 5.10   Show loop for pocketing the first section of the song before the drums come in.

Use Cmd-] (Mac) or Ctrl-] (PC) to zoom in close to the first bass note under the hat loop. By looking closely you can see that the bass transient is virtually sitting dead on top of the click transient.

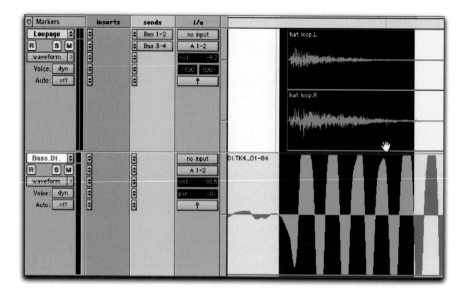

Figure 5.11   Too on top. A bass sitting right on the beat feels a bit rushed.

While this might work fine if the song is supposed to start with a bang, in this more smooth intro it actually feels a bit too rushed, or on top of the beat. While every major editor may have his or her own opinions about timing and feel, this particular song needs instrumentation that provides a laid-back, relaxed atmosphere, without getting sloppy.

Since we have already pocketed the drums to this loop, and they're 'leading the charge' of the rest of the instruments, we'll be looking to pocket all of the instruments about 3–8 milliseconds behind their beat. This will give us a fluid feel, but with solid presence.

A look at the pocket: Where does the bass wave begin and where should it end?

Now look at the first two cycles of the bass waveform. Generally, the first few visible cycles of a bass wave are actually a finger or pick hitting the string. The real start of the note often comes about four cycles in. Therefore, if the very first bass transient starts right on top of the loop transient, it will actually feel a bit early.

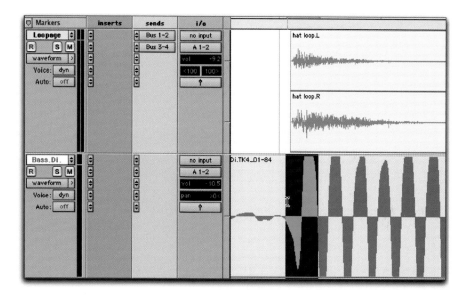

Figure 5.12 First few cycles of some basses are often just the string hit. The note starts later.

There are two basic ways to fix this sort of pocketing puzzle. The first is to use your Fade tool to simply add a small fade to the very beginning of the bass note. Start by selecting your Smart tool with the selection button on your toolbar.

Figure 5.13 The Smart tool: The only tool you'll ever need.

Separate, Fade, and Nudge: A simple bass pocket

Next, we need to separate this note into its own region. We're also going to mute the bass guitar parts that occur before the loop starts so that the song has a new dynamic feel to come in with the loop.

Start by hovering the Smart tool over the top half of the region right where the transient of the bass note starts to build. Your Smart tool will become the 'I' of the Selector tool. Now type Cmd-E (Mac) or Ctrl-E (PC) to separate it into its own region.

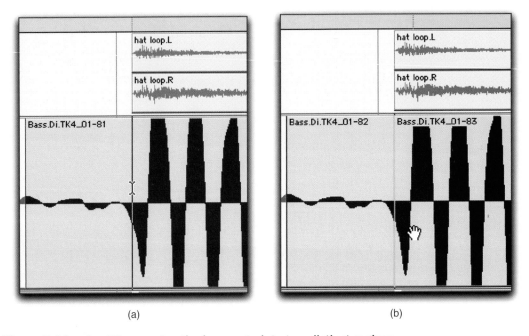

(a)   (b)

Figure 5.14   Cmd-E separates the bass note into two distinct regions.

Next, hover your mouse over the top left half of the bass note, and you will see it turn into a small square with a fade through it. Simply click and drag to the right a bit to create a short fade over the finger strike sound. This will effectively eliminate the strike of the pick or finger, allowing the note to pop through just slightly behind the beat, for a more relaxed vibe.

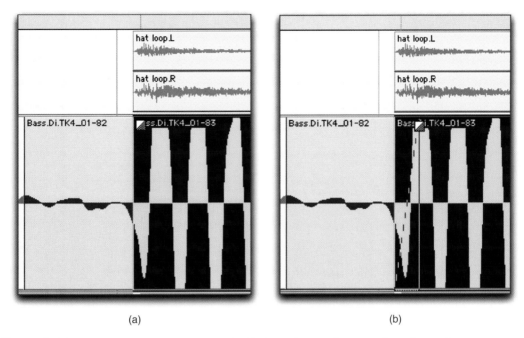

(a)   (b)

Figure 5.15   The Smart tool becomes the Fade tool over the top edge of a region.

The other approach is to simply use Pro Tools to move the note a few milliseconds later in time.

Start by making sure your Nudge value is set to Min:Secs under the Nudge submenu in the toolbar.

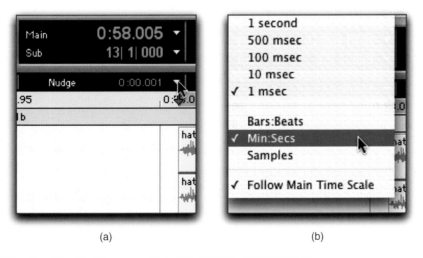

(a)   (b)

Figure 5.16   Use the Nudge menu for pocketing to milliseconds.

Some people like to use values of 5–10 ms for creating special effects, but this approach doesn't work as well for pocketing. Set it to a simple 1-millisecond value, select the bass region that you just added the fade to, and then use the plus (+) and minus (−) keys on your keyboard to slip the bass track about 4 ms later in time.

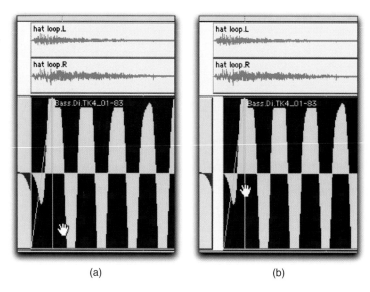

(a)   (b)

Figure 5.17   Nudging or dragging the bass 8 ms behind the beat.

Two schools of thought for pocketing bass

It's here that we run into two different schools of thought on pocketing in general, and bass in particular. Notice that when you nudged that first note to 4 ms later in time, the rest of the bass track followed with it. Now, the entire bass track, whether it was originally on the beat or not, is now 4 ms later in time. As a result, some engineers prefer to use a piece of software like Beat Detective (that we saw earlier with the drum kit pocketing) to first cut up all of the notes into individual regions, and then simply pocket them one region/note at a time. For some styles of playing this may work OK and can occasionally save a significant amount of time. For the best of us out there who are going to put the whole track under a microscope anyway, and go through it one transient at a time, Beat Detective on instrument tracks can often be more hassle than it's worth. If it misses notes, or the threshold is set wrong, you will have to go back through the track to fix all of the little problems it creates, and you still won't have a track that was pocketed by ear.

The best approach in my opinion is to simply go through the song, quickly pocket your notes both audibly and visually, and presto-chango, you're tight as a drum, er, bass.

Run, Spot, run: Spot mode comes to the rescue when nothing else will

Another advantage to simply editing/moving the whole track is for when you have a brain spasm in the middle of the track, and can't really remember or necessarily hear whether you've moved the most recent note by 2 ms or 30 ms. In this situation you can simply change over to Spot mode (of course, making sure Auto-Spot is turned 'off' under the Operations menu), select the track being pocketed, and click the up arrow next to 'Original Time Stamp' to move the rest of the unpocketed track region back to its original time location. This way, you can essentially get a third grade 'do-over' and continue on in your editing state of bliss.

Figure 5.18

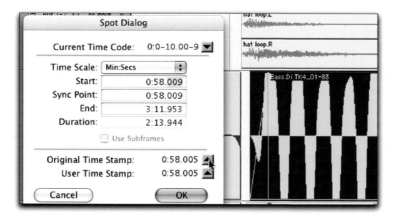

Figure 5.19 Spot Dialog for moving regions back when you make a mistake.

For now, we can stay in Slip mode and go on with our lives.

A closer look at the bass

The next issue is one of consistency in your pocketing. Since the whole point is to bring the tracks to a tighter, more centered performance, an instrument like the bass guitar must be edited to a consistent standard. For example, look at the bass waveform in Figure 5.20.

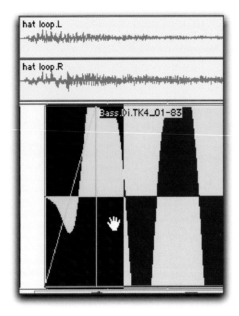

Figure 5.20 The string sound – not used for pocketing.

We have previously discussed how this initial cycle or two is often the sound of the string being struck, and that the note itself doesn't tend to sound until a few cycles later (on this more 'wubby' sounding bass guitar at any rate). Since this bass guitar will generally have a consistent sound through the whole track, we have decided that the start of our note (the piece that we will be lining up against our other tracks) will generally be the third or fourth cycle of the note.

That said, we then need to consistently pocket this bass guitar against that portion of the wave as the start of the note. In other words, if we are sloppy about our pocket, sometimes lining up later in the wave or right before the wave starts, then we will have an inconsistent sound and will have wasted our time with this whole process. So, it really doesn't matter as much *where* in the sound you choose to pocket against, as long as you're consistent about it.

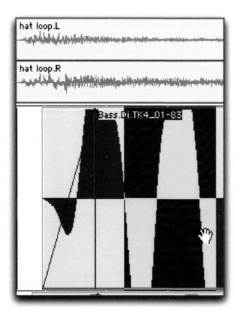

Figure 5.21 The actual start of the note that should be used for pocketing.

We've chosen to start our notes and edit them to start around three cycles in.

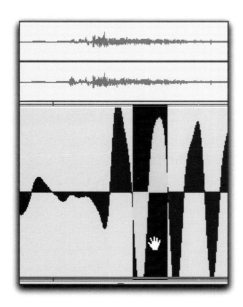

Figure 5.22 The real start of the note. Pocket this area 5–8 ms after the loop transient.

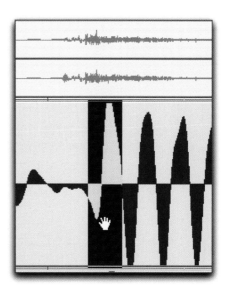

Figure 5.23 Probably the strike of the pick or finger on the string. Pocketing to this may make the bass sound early.

187

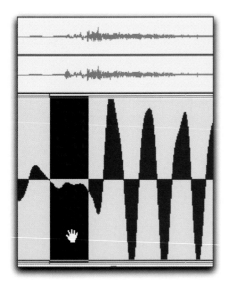

Figure 5.24 No good. Using this as the start of your note will make your bass sound late.

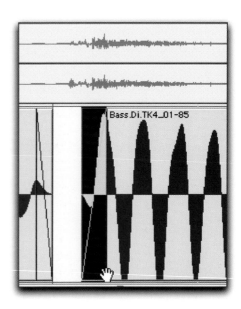

Figure 5.25 A good edit. It fades out the garbage end of the previous edit and smoothes the string pick at the beginning of this note.

Pre-rolls, post-rolls, and solos: Repeat that five times fast

The last step to set up is your pre-roll, post-roll, and solos (try saying that five times fast). Cmd-K (Mac) or Ctrl-K (PC) will toggle on pre- and post-roll for your playback, and will allow us to make our edits, and then simply hit the Space bar to play back a few seconds before and a few seconds after our edit.

Figure 5.26 Setting a good pre-roll time.

This will let us hear our edit in an audible context, so we can make sure it works in the rest of the mix. For our bass guitar, since the notes are long and we're not going to look at every 16th note, we can use a longer pre-roll of 5–6 seconds. Now, whenever we hit Play the track will play back 6 seconds before our edit and to 6 seconds after (or until we hit Stop). Finally, select the Solo button on your kick, snare, bass, and loop tracks, and let's start the pocket!

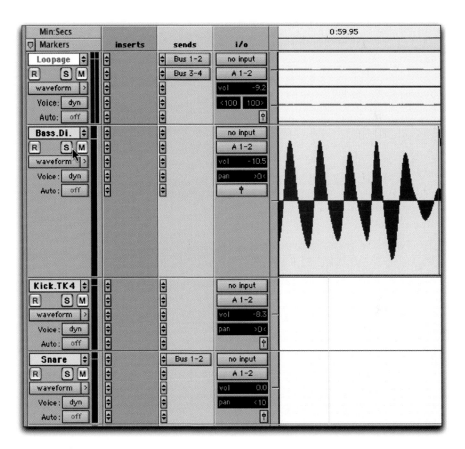

Figure 5.27   Soloing only what you need to hear.

Now that you have everything set up, zoom in on the next bass note located around 1:01.257. Use your Selector tool to select right before it and hit Play. Notice that while the transient is hitting right with the loop, the sound is just a little 'fwubby'. A quick way of handling this is to just select and delete those few offending cycles.

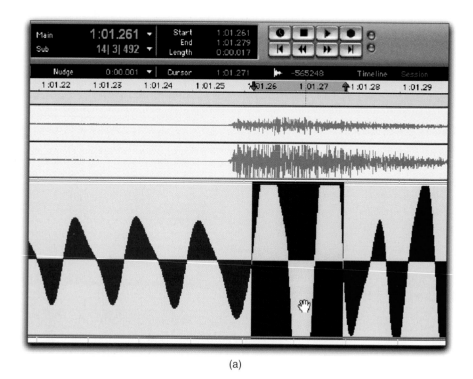

(a)

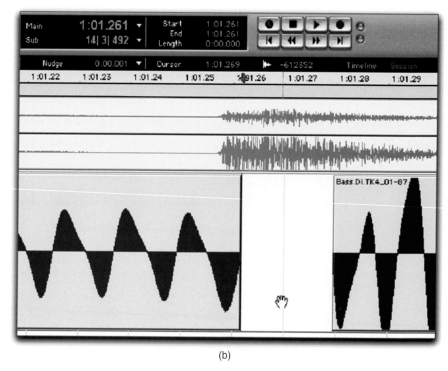

(b)

Figure 5.28   Deleting audible bumps/noise.

Now select your track, slide it back (earlier) a few milliseconds to fill in the gap, and apply a quick crossfade. Voila! De-fwubbified!

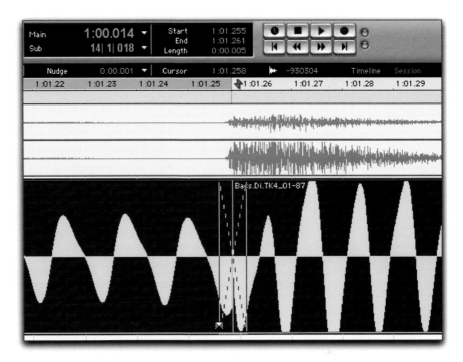

Figure 5.29   A quick and dirty crossfade.

Double trouble, double bass transients, and how to fix them

Of course, you didn't think it'd be that easy did you? Well, depending on the note, it might be or it might not. For example, look at the crossfade we created on the wave in Figure 5.30.

Notice that even with the crossfade it has actually made a bit of a double transient that can regularly sound like a small 'click' or 'pop' – which, as they say, is not good. The solution to this is to actually slide your note around until you can get one of the zero crossings to match up with the edge of the previous region. To do this, change your Nudge submenu to Samples rather than Min:Secs, and set it to 1 or 2 samples.

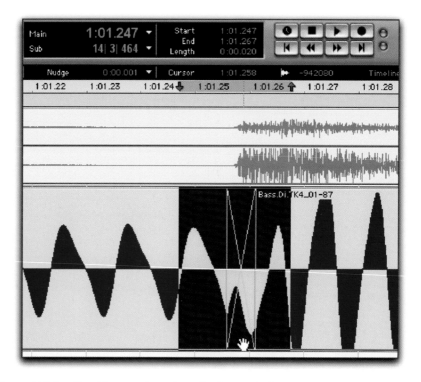

Figure 5.30 Double transient from bad crossfade.

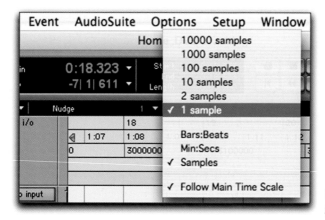

Figure 5.31
Setting to samples.

Next, select and delete the bad crossfade, select the later region, and try to match the beginning of its positive cycle with the end of a negative cycle on the previous region. Now, zoom into the sample level, and use your plus and minus keys on the keyboard to nudge the waveforms until they meet at a zero crossing.

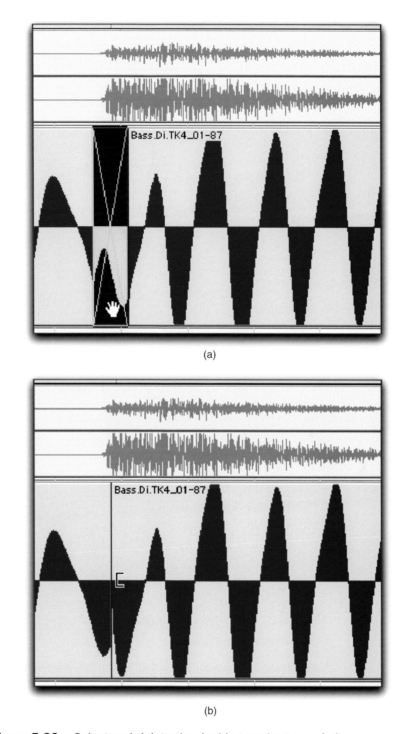

(a)

(b)

Figure 5.32 Select and delete the double transient crossfade.

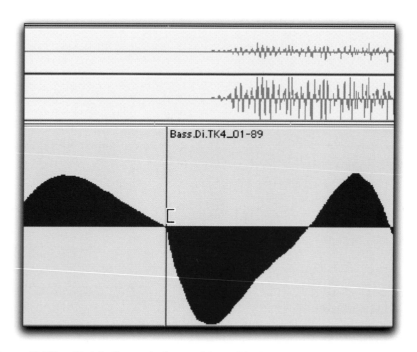

Figure 5.33   Match the end of a positive with the beginning of a negative cycle.

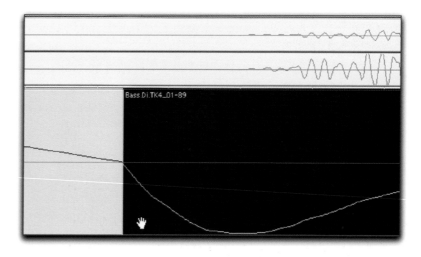

Figure 5.34   Matching waveform edit at the sample level.

Editing without crossfades: The 'nudge at the sample level' routine

Zoom in to the sample level and try to match up the zero crossing points of the two waveforms. This will often save you from having to create a crossfade, and can make even the toughest edit work just fine.

While the time and energy it takes to make this happen is obviously more intensive than just applying a quick crossfade and being done with it, you will occasionally have an edit that just won't be easily worked, and will need to be nudged at the sample level (rather than just milliseconds) to find a happy medium and eliminate the 'pop'!

> **Problems to look out for**
> Tab to Transients is a great feature in Pro Tools and other software that allows you to quickly maneuver through tracks with big transients. However, on tracks with large, sustaining waveforms like a bass guitar, it's not always the most accurate at finding every subtle note in the middle of a performance. As a result, you need to keep your eyes *and* ears open while you're pocketing to make sure that those subtle notes don't get skipped over.

Now, if you're using both your eyes and ears as you play back this next note, you'll realize that it is in fact not one note, but two. However, if you were tabbing along with Tab to Transients, Pro Tools likely did not catch that second note because there was very little amplitude difference between the second and the first note. Before moving on, pocket the second note so we can use it to learn another handy skill.

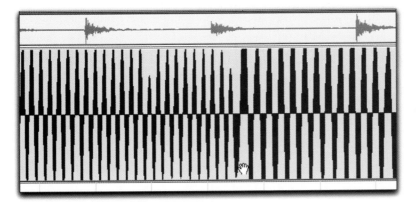

Figure 5.35 The hand shows where the missed second note occurs. Notice how far behind the loop beat it is. It's just begging for you to pocket it!

It should look something like Figure 5.36 when you're done.

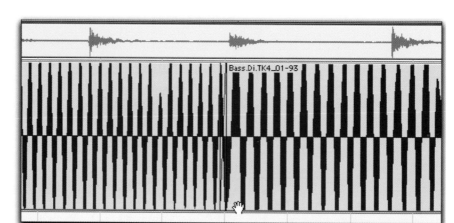

Figure 5.36   Notice how the transient of the bass note is pocketed against the loop.

Now that the hidden second note is discovered and pocketed, another situation where you'll use the old 'nudge at the sample level' routine will be when you have to fix a bass note by extending it to fill in a gap created by either a missed note or a tough edit. Skip over to the note that begins around 1:05.264. If you look at the end of it, we see some fret noise occurring between it and the next note.

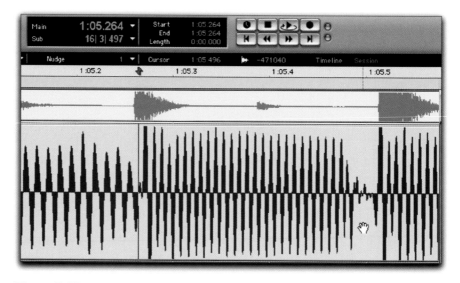

Figure 5.37   The hand points at the offending fret noise.

Moving the regions, filling in the holes

The previous solution called for simply deleting the noise, pocketing this note and the next note, and fading the remaining edges, which would look something like Figure 5.38.

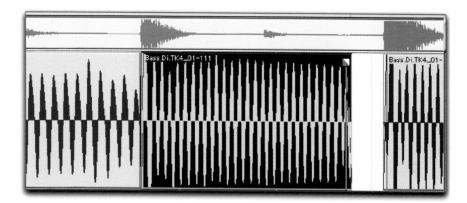

Figure 5.38 Noise deleted, notes pocketed, fades created, mission accomplished.

However, if it would sound better for the note to sustain through the entire section, we need to somehow extend it to fill the gap.

In order to do this, go ahead and delete the noise, pocket the next note, and then use Cmd-E (Mac) or Ctrl-E (PC) to separate the region somewhere in the sustaining middle of the first bass note.

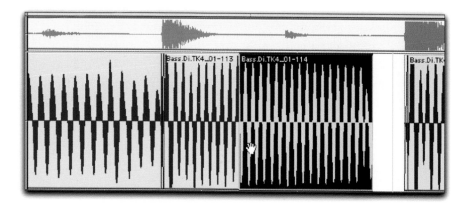

Figure 5.39 The region is split using Cmd-E.

Then drag the end of the region to the end of the gap.

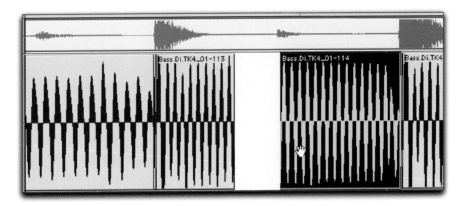

Figure 5.40   The end is nudged later to fill the gap.

Now, using your Trim tool, select the left edge of that region and extend it back to fill in the middle section of the note.

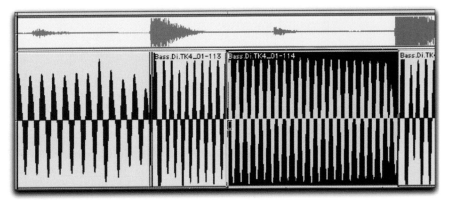

Figure 5.41   The regular Trim tool is used to extend the note back to fill in the gap.

Zooming in close to your edit point, use your Nudge at samples or milliseconds to line up the curves of the waveform until you either find a perfect zero crossing or you line up two edges that perfectly match.

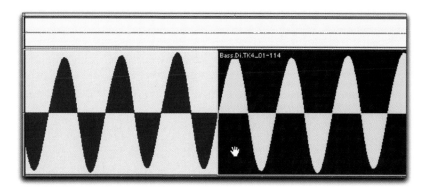

Figure 5.42 Zoom in close to line up the waveform cycles.

If you align this just right, it is possible that you can even get away without using a crossfade, and will have magically extended your bass note to fill in the space. As they say on those Guiness commercials, 'Brilliant'!

Overcoming the beast of fret noise

Fret noise – the bliss of the lead electric is the blight of acoustic and bass guitar players everywhere. On a truly professional record, like any other professional release (movies, television, etc.), there generally isn't anything in the end release that isn't intentionally put there. Fret noise, pick noise, string burps, and real burps all need to be edited out of the final product so they don't distract from the intended concept, message, or delivery of the song. Often, simply editing out these little screeches and bloops and adding a slight fade to the audio before and after is both preferable to the noises themselves and enough to add a touch of dynamics to the sound. In addition, while these little sonic quandaries may seem somewhat insignificant in the initial stages of cutting a song, as you start to mix, EQ, compress, and effect your tracks, these little sounds will start to make themselves collectively known by affecting the feel of your track. For example, compression on the bass or the mix bus towards the end of a mix will make fret noise more audible, but you may not notice it because your ears are tired and accustomed to the sound. At the same time, if fret noise before a note is even slightly ahead of the beat, it will have a tendency to make the whole note feel early.

Like we saw before, one way to eliminate these problems is to simply delete and fade. However, in a sustained instrument like the bass guitar, simply cutting out the fret noise at the end of a phrase can sometimes leave a conspicuous hole or glaring gap in the mix. For instance, examine the fret noise in Figure 5.44. This almost sounds like a mistake in the track, but to simply select and delete it leaves a rather obvious hole in the sound. To solve this dilemma, we reach into our magic hat for the same conjuring trick we used in the drum editing: Time compression and expansion.

Time compression and expansion: Using the TCE tool to fix bass gaps

Let's first try to fix this fret-noisified bass note using Digi's TCE tool. First, pocket the transient of the bass like we've already seen against the transient of the loop track.

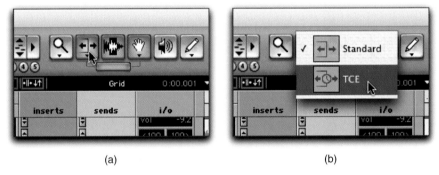

(a)   (b)

Figure 5.43   Click and hold down the Trim tool to get the time compression expansion (TCE) version.

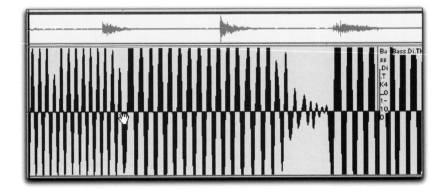

Figure 5.44   Unpocketed.

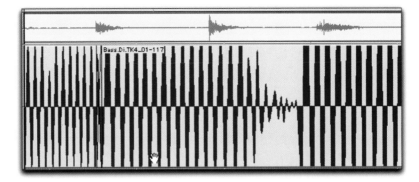

Figure 5.45   Pocketed.

Now, you can see the cluster of low-amplitude waves that is our fret noise at the end of the note. Select and delete this fret noise, pocket the next note, and then we'll fill in the gap created by the discarded noise.

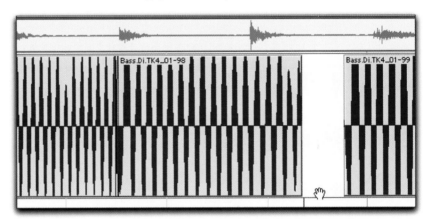

Figure 5.46   Noise deleted.

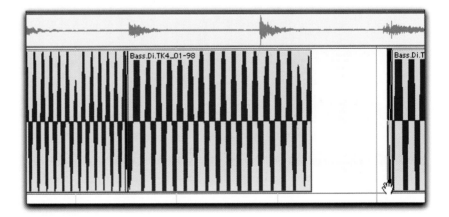

Figure 5.47   The next note pocketed.

Previously, we copied a bit of the middle of the note to help extend it through the break. This is one way of filling a gap that works best on notes that have long sustaining sections in the middle. For shorter notes, we now make a similar edit during the sustaining part of the bass sound by using our Selector tool and hitting Cmd-E (Mac) or Ctrl-E (PC) to split the region.

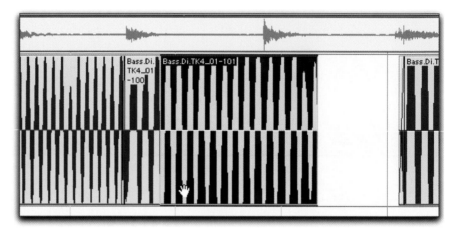

Figure 5.48   Note the second half of the note is now its own region.

Now, after selecting the TCE version of your Trim tool, simply select the end of the note and stretch it to the right to fill in the gap.

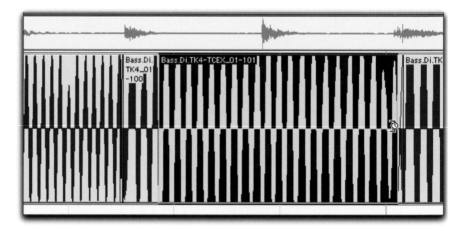

Figure 5.49   Note the Trim tool with the small clock on the right of the time-stretched selection. This has essentially stretched the end of this note out to fill in the gap. Notice also the letters TCEX added to the selected region's name. This denotes that this region has been affected by the TCE tool.

You may need to apply a very small crossfade to the edit point between the original transient and the time-expanded section, but frequently this will not be the case. If you've used a good TCE tool and the expansion wasn't too extreme, this will sit just fine in the track and fill in the empty space. While it's true that you won't run into this kind of random fret noise in every situation, knowing what to do and how to properly get rid of it is key.

When nothing else works – try Copy and Paste

But what about the noises that happen in the middle of a note: The flubbed picks, the double strikes, and the accidental plucks? Here's where good old-fashioned Copy and Paste comes to our rescue. Take a close listen to the bass guitar around time 1:05.777 to 1:06 on your timeline.

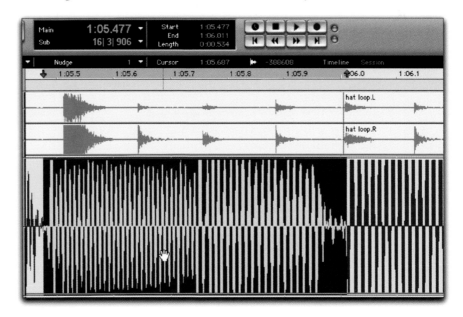

Figure 5.50 An overly fwubby note.

It's not huge, but you can hear the player mildly tweaks a note that wasn't caught during tracking. While this sort of thing can slip by in the heat of a rocking tracking session, you can guarantee the bass player will hear it and point it out every time he hears it after the final mix is pressed, packaged, and shipped. No problem. With a quick jump back to the previous riff around 1:01.500, we can find the same F# that we just time expanded and pocketed a moment ago.

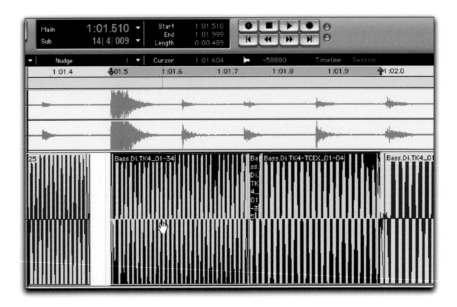

Figure 5.51 Copying the previous good F#.

Shift-click both pieces to select them. Cmd-C (Mac) or Ctrl-C (PC) copies it to our clipboard. Select our less fantastic F# and use Cmd-V (Mac) or Ctrl-V (PC) to paste it over the top!

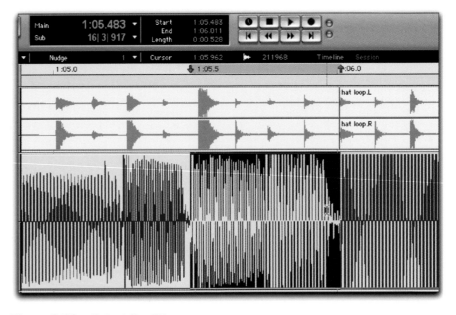

Figure 5.52 Select the F#.

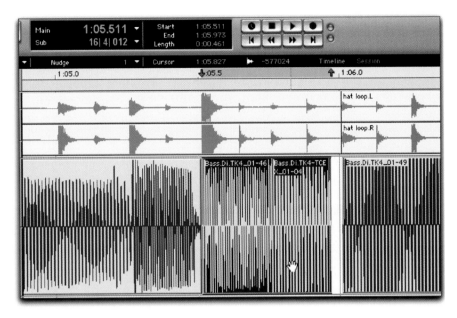

Figure 5.53   Pasting the good note over the bad.

A few milliseconds nudging to pocket it against the loop, and lo and behold –
the mistake never happened. You're getting to be too fast for your own good.

Using your eyes *and* your ears: A wrap-up

All of these bass tips are useful for either fixing a challenged performance
or as creative tools for an already great one. Even if you're working with a
great bassist like the one on this session, these tools really become more of
a 'mixing' tool than 'fixing' tool. For example, if the player injects a creative 'pop' at the top of a note, you can pocket the pop as the front of the
note, rather than deleting it out of hand. On the other hand, if fret noise
is prominent in the mix, it can make the bass player seem to be early, even
if, in fact, they were not.

Now that you know the simple tricks, make a pass through the bass guitar
track. First, just focus on going through and pocketing the notes against the
kick and snare, or against the loop during the breakdowns. After your
notes are pocketed, you can take the time to go back through and terminate
all of the little sounds that will contribute to a less-than-stellar bass track.

205

Always remember, the whole reason we're doing this pocketing thing is to make the performance sound its absolute best. When you're going on a date, you put your best foot forward, and tailor the approach to fit the situation. Pocketing is no different. If the bass needs more attack, sometimes space before the notes can actually make them pop out even more. If an accidental note actually works well in the song and contributes positively to the overall vibe, let it ride. Don't allow yourself to be sucked into the idea that 'not perfect' is 'not good', and always remember, waveforms are not your ultimate guide, your ears are. If it sounds good, it *is good*.

Finally, we'll do the same thing to the bass guitar that we did to the drums and percussion tracks by creating a new playlist, consolidating our bass edits to one region/file, and renaming both it and the track to be Bass_M, indicating our master bass track.

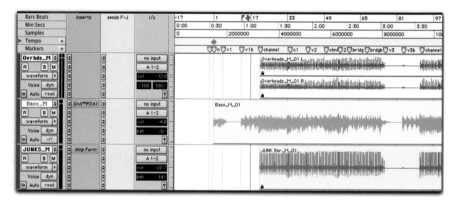

Figure 5.54   Our master bass track consolidated to a new playlist and renamed.

This leaves us ready to move on to our acoustic guitar.

Locking up the acoustic tracks

A bad analogy: Building our pocket

Now that your low-end is doing some serious booty shaking, it's time to examine the filling and icing on our delicious little musical confection. If you want to consider the drums and percussion as our base-level ingredients, and the bass guitar as the substance of the mix, then our acoustic and electric guitars must glide silkily over the top and provide an attractive backdrop for our candles and little plastic figurines – the vocals. Did I mention I'm terrible at making analogies? Yeah, I am.

In case you haven't yet detected the pattern, after we initially spent so much time pocketing the drum tracks to the loop/click with all of our tabbing and separating, time compression, and special crossfades, we have moved on to pocket all of our instruments to … the drums! Betcha didn't see that coming. We started with pocketing the drums and loop, and as we moved to percussion we had our percussion, drums and loop, and on to bass, where we utilized our bass, drums, and some percussion and loop.

Essentially, as we move up the instrument pocketing chain, we now have the freedom to leave more and more tracks audibly in the mix so that we're constantly building on a fully pocketed track. If we just listened to the acoustic and drums as we pocketed, it may end up working perfectly with the drums, but have parts that clash with the bass. If we moved on to the electrics and just pocketed them against the drums, without also hearing the acoustic and bass in the mix, we would run into the same problem. So, as we move through pocketing all of the tracks, we're going to continually add tracks to our mix so we can hear our song build up over time.

Obviously this doesn't work in reverse because, once we've pocketed our drums, the rest of our tracks will be out of time until we pocket them, so consider making it all the way to the acoustic guitars as a sort of milestone on your way to multi platinum credits!

Setting up our acoustic pocketing Edit window

Starting with the acoustic guitars, our approach is going to utilize those rocking great drum tracks we've grown to know and love so much. Just like before, let's first set up our screen to view only the things we need to see. This time, however, leave your previously pocketed tracks in the mix at a modest level so we can really start to feel our track as it grows and takes shape.

Go back to your Show/Hide box and show the acoustic guitar track while hiding any percussion tracks that we don't need to be looking at right now. You can leave them in the mix audibly, but we don't need them cluttering up the screen.

Figure 6.1 **Show your acoustic guitar and move it between the kick and snare tracks.**

Pocketing without drums: How to deal with it

We've already seen that during the intro of the song where the acoustic track begins, there are no drums. The drums don't come in until the channel of the song leading into the first chorus. What to do? First things first, we should probably take a listen to the track to see what else is going on with it. If the acoustic has any automation on it, turn it to 'off' and turn up the track so we can really hear it in the mix.

(a)

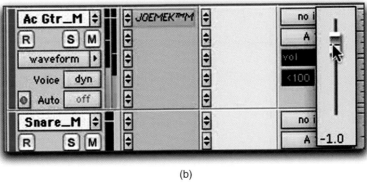

(b)

Figure 6.2 Turning off automation and turning up the guitar.

Well, for now we can assume that the hat loop track will be staying in for the remainder of the mix, and since it's the only sort of timing reference we have here at the top of the track, let's go ahead and pocket our acoustic guitar to that. We're almost as resourceful as MacGyver now aren't we? Now where did we put that paper clip and duct tape?

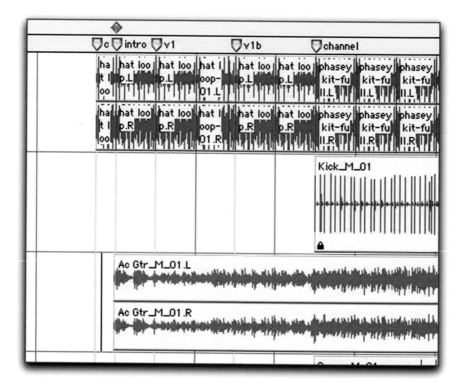

Figure 6.3   Hmmm. No drums to pocket the intro against – let's use the hat loop.

Start by temporarily moving the acoustic guitar up directly under the loop track, and let's resize it to 'large' rather than 'medium'.

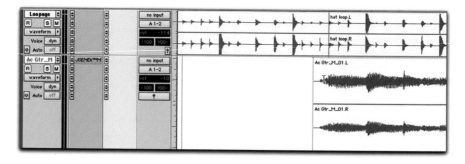

Figure 6.4   Moving our acoustic guitar and resizing it for pocketing the intro.

Dealing with raked chords and trusting musicians

The very first chord he plays is what I would call a raked chord. It has a really drawn out transient. As a result, we're really not going to try to look for a specific transient to pocket against. Rather, we're going to just make sure it feels okay in the context of the track and against the loop.

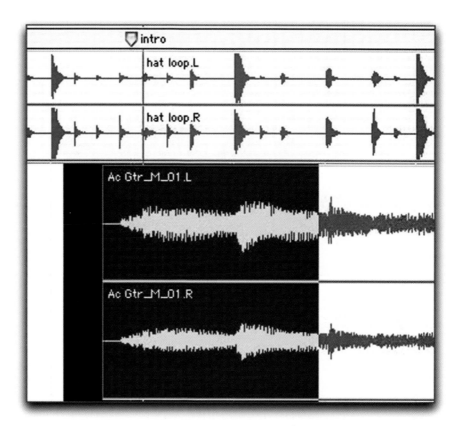

Figure 6.5 The raked intro chord – where's the transient?

This brings up a point that we're going to start running into more and more as we start working with our guitar tracks, but would also apply if we were working with piano, strings, etc. If you find yourself on a note where you cannot tell if it's early or late by looking at it or listening to it, then trust the musicianship of the player and leave it where it's at. Spot it back to its original location if need be, and just make sure it maintains its relative spot to your other pocketed tracks. All of the musicians that played on this song were incredible, so we can trust their musicianship. This is not a rigid science that says everything *must* be in a certain place, it's an extension of the art of mixing that just works with the musicians to create a great performance.

Now, if you have a really obvious transient like in Figure 6.6, go ahead and pocket it. Otherwise, don't stress about it as long as it feels good.

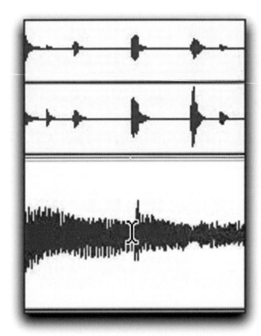

Figure 6.6   Now that's a good transient.

Since the first chord feels pretty good where it is, move on to the second note around 1|2 that does have a decent transient for pocketing purposes. Also, vertically zoom in once or twice to blow up the waveform to a usable size.

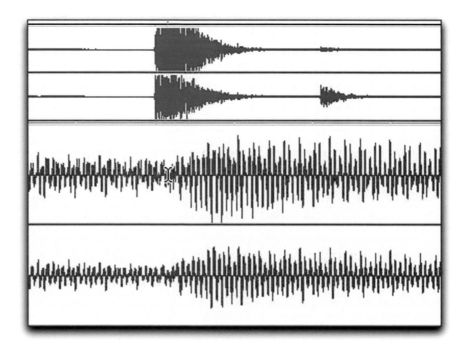

Figure 6.7 The second acoustic chord is a little too far behind the beat.

Understanding the acoustic waveform and how to pocket it

Obviously acoustic guitar is not going to be as neatly defined as a drum or even a typical bass guitar waveform. It has a lot more to look at, with the waveforms being a lot closer together, and it's sometimes difficult to tell where one note ends and another note begins. In Figure 6.8 I've blown it up a bit so you can see that my Selector tool is showing the lowest point between the first note and the second. If you look at the left side you will see the waveforms decaying over time, and right about the insertion point it starts to grow into the full transient of the next chord. Go ahead and separate the regions and pocket your second chord against the hat loop above it. It should go from starting at around 1|2|009 to 1|2|025.

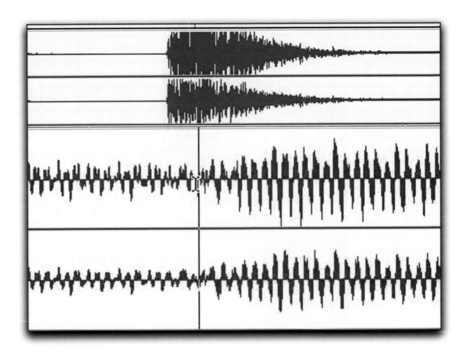

Figure 6.8   The transition from the first chord to the second.

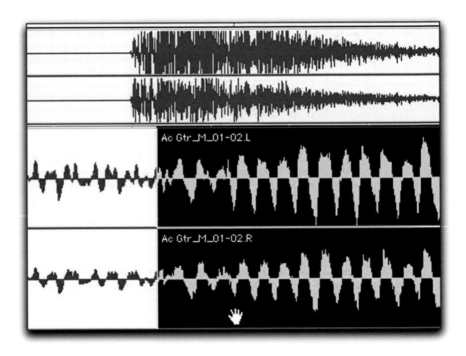

Figure 6.9   The cut is made and we drag the region back to pocket against the loop track.

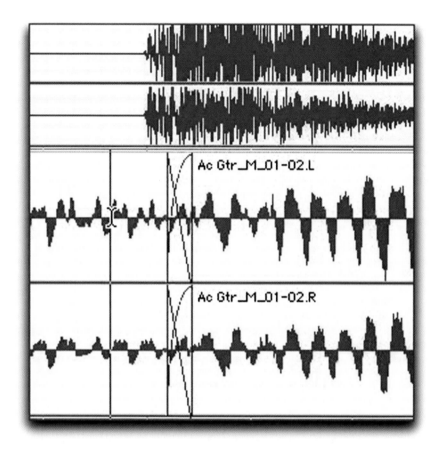

Figure 6.10 Add a crossfade and take a listen.

Sounds good. OK, move on to the next note around bar 1, beat 3 and repeat the process. Find the lowest point before the waveform starts to increase again, insert a cut, slide the region into the pocket with the hat, listen to it in the track, add your crossfade, and done.

215

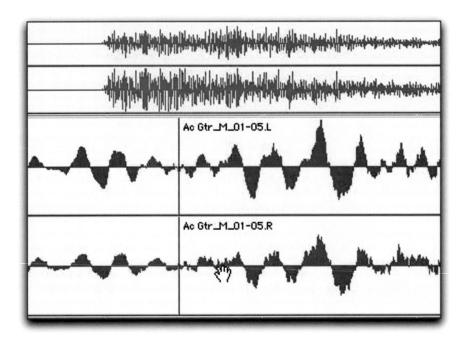

Figure 6.11   Cutting the next note.

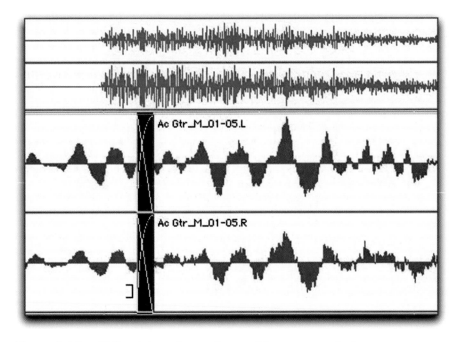

Figure 6.12   Sliding it into the pocket, making our crossfade.

Now we're cooking with gas. The next note around 1|3|457 looks like it might be a bit of a special case. Take a look at the selected area in Figure 6.13.

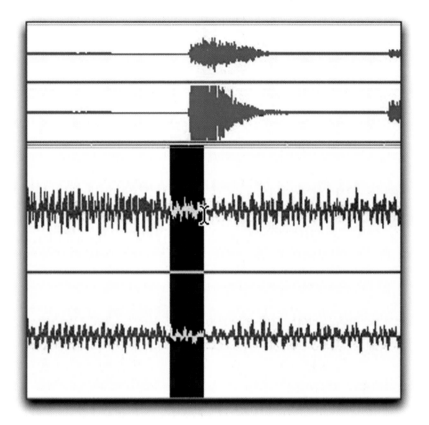

Figure 6.13 This fuzzy area looks like a pick or fingernail sound leading into the next note.

The fuzzy area noted in the picture looks suspiciously like a pick or fingernail raking the strings before leading into the next note. Often, if a musician does this we end up with the note sounding like it's coming in early due to the pick noise, even though the full note doesn't ring out until a little later. The only way to tell for sure is to turn your pre-roll back on for a couple of bars and listen to it.

Figure 6.14   Turning pre-roll back on.

In this particular place, it still feels OK to me, so we're going to leave it alone. However, this is a pretty common problem when working with acoustic guitar tracks, so keep your eyes open for it.

The next note hits around 1|4|000 and when we zoom in close it already looks to be pretty in time with the hat loop. Since he's not on top of the beat and it feels good in the track, we're going to go ahead and leave it alone for now. Remember, we don't need to edit every note just for the sake of editing every note. If it sounds good, it *is* good.

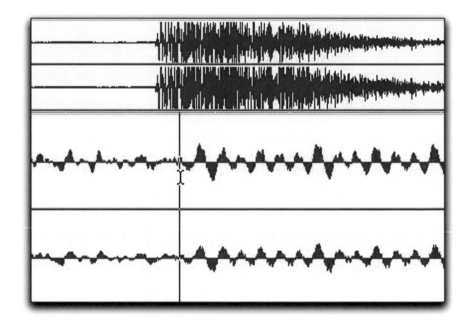

Figure 6.15   The next note at 1|4|000 feels good. Let's leave it alone.

Looks like we spoke a little too soon. The very next note at 1|4|457 has a big rake right in front of it, and it puts the note itself a little too far behind the beat.

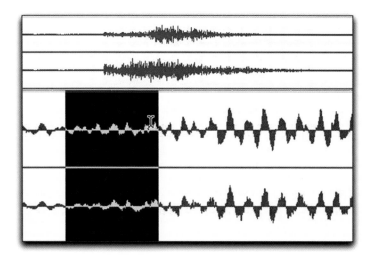

Figure 6.16   A big rake coming into the note. This puts the note too late.

My solution in this situation will be to just put a cut in front of the note itself and pocket it over the latter half of the rake. Since the raking sound is so long we won't completely cover it, but just pocket the note like we have previously, shortening the sound of the rake. Finally, add your cross-fade to cover the edit, and you're ready to go. Give it a listen to make sure the timing and edit work before moving on.

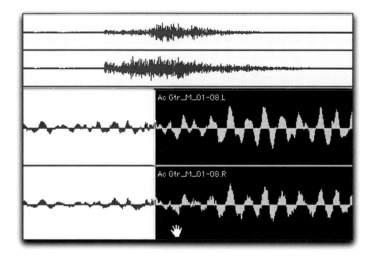

Figure 6.17   Separate just the note.

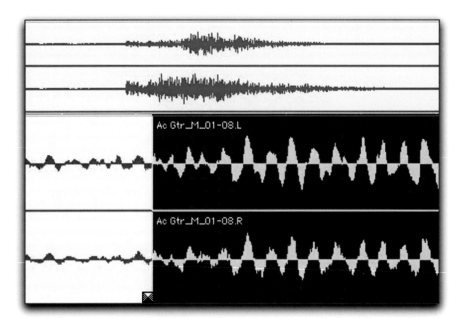

Figure 6.18   Pocket it against the hat transient, shortening the rake sound.

Pocketing has left our track early: Pitch 'n Time to the rescue again

Very good. When we get to the end of the next long-held note around 2|4|677, we start to notice a problem. As we've been pocketing the top of this track we have been consistently moving the acoustic region a little earlier with every edit. As a result, the end of the held note is now ending quite a bit early. This is not just a problem for pocketing the next note a little later, because if we do that, there will be a gap between the end of the held note and the next chord.

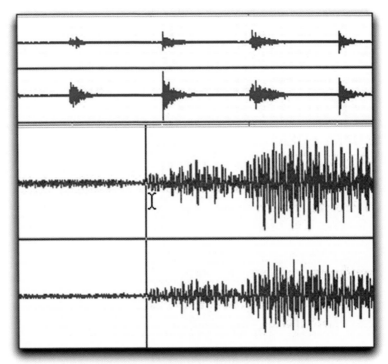

Figure 6.19 The end of the held note is ending too early.

For further proof that all of our notes are ending up too early, just look at the next few notes after our held note. Notice that all of their transients are now consistently ahead of the hat loop.

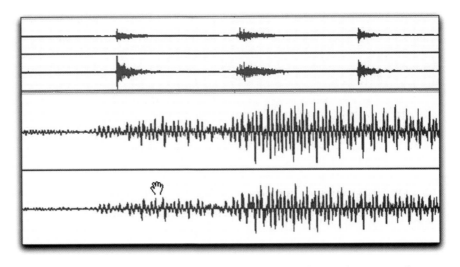

Figure 6.20 Our pocketing has put the latter chords starting too early.

The quickest way to resolve this is to drop an edit in at the end of the held note and then either nudge the later region into time or just use our Spot mode to put it back where it started. Let's use Spot mode for now.

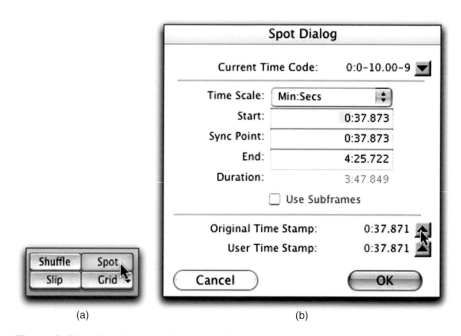

(a)   (b)

Figure 6.21 Use Spot mode to put the track back where it started.

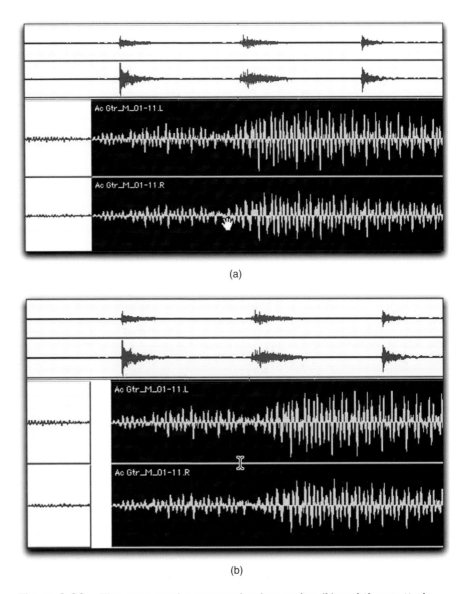

(a)

(b)

Figure 6.22 The unspotted, separated guitar region (L) and the spotted guitar region (R).

Now, when we examine both of the next notes in their current spotted position, we see that the second note is right about where it needs to be, but the first note is actually still ahead of the loop track.

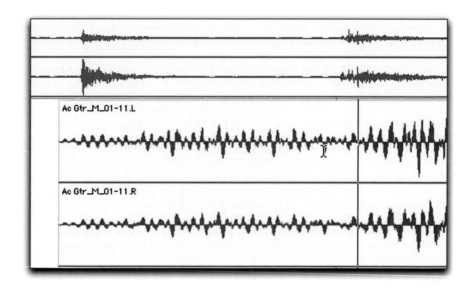

Figure 6.23 The second note (right) is about where it needs to be, but the first note (left) is actually starting before the loop.

Let's fix this by placing a cut at the head of our second note, and sliding our first note to the right to fit into its own pocket with the loop.

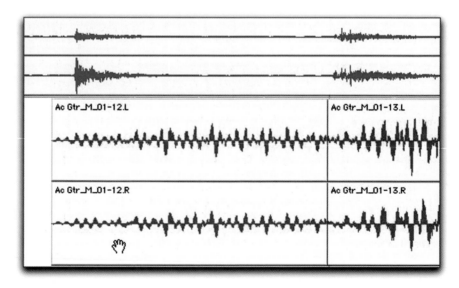

Figure 6.24 Cutting between the notes.

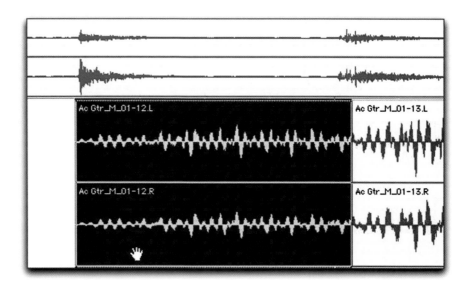

Figure 6.25 Pocketing the first note.

This puts the acoustic note a little over the top of our second note, so let's just use our Trim tool to trim off the end of our first note to reveal a little more of our second note, add a crossfade, and then take a listen.

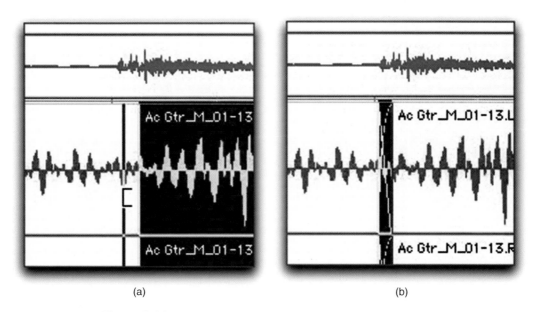

(a)   (b)

Figure 6.26 Trimming back to get the top of our second note and crossfading.

While it's starting to look like we've got multiple pieces sitting all over the place, my goal is to line up the main downbeat (in this case the second note), and then just move the quieter transients (i.e. the first note) to where it needs to be.

Well, now we need to look at that hole we left between the held note and the ones we've just been editing. Again, we can't just trim the right region back to fill in the space, because then we have the same audio playing before the crossfade and after the crossfade, which will be audible just about every time, and make you sound like a sloppy editor. The key is going to be to go back to our friend the Time Compression/Expansion tool.

Like we mentioned earlier, we have been using the Serato Pitch 'n Time plug-in for our time compression/expansion, and the TCE'd regions sound really good. If you are having to use Digi's default TCE plug-in, you may have to massage these pieces a bit more, or have them down in the mix just a bit to cover up the bubbly sound often produced. Either way, it should still work well enough for you to get a handle on TCE'ing the necessary parts.

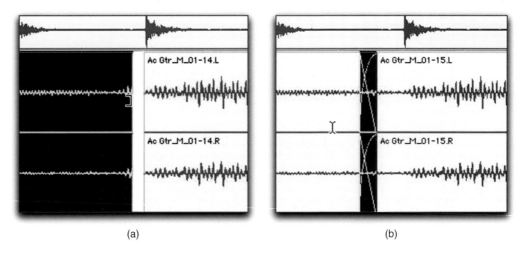

(a)   (b)

Figure 6.27 The problem with a gap in the audio again. Don't just trim the right region back – this will leave the same audio playing before and after the crossfade and will be audible.

Start by trimming the held note out past the gap until you see the transient for the next region appear right next to itself.

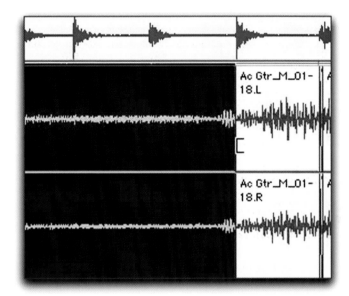

Figure 6.28 Trimming the held note over the gap until the next region's transient appears.

Now zoom out and go back towards the meat of the transient for the held guitar chord. Insert a quick cut to separate the main transient of the wave-form from the decay. Around 1|4|916 should do it.

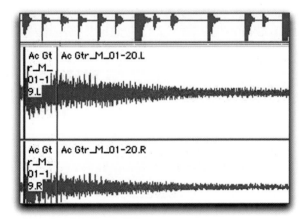

Figure 6.29 Separate the decay from the transient.

Now switch back to your TCE Trim tool, zoom in to the end of the waveform, and stretch it until the transient at the end of your region overlaps the pocketed later version of itself.

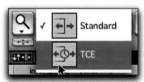

Figure 6.30   Switching back to the TCE tool.

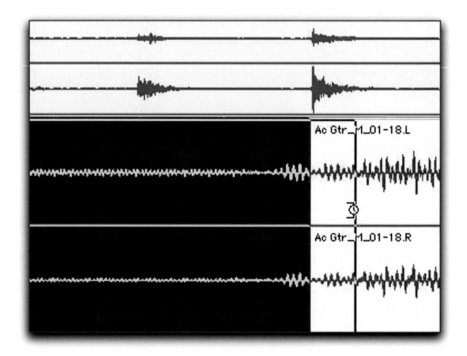

Figure 6.31   Time stretching the next note transient over its pocketed version.

Now, switch back to the standard Trim tool, and trim the already pocketed region back over the time-stretched version of itself. Crossfade the time-stretched region where you cut it, and between the next note, and play the section back to make sure the edit works. Like before, it may ask you to adjust the bounds of any crossfade between a time-stretched and non-time-stretched region – just tell it to Adjust Bounds and it will be fine.

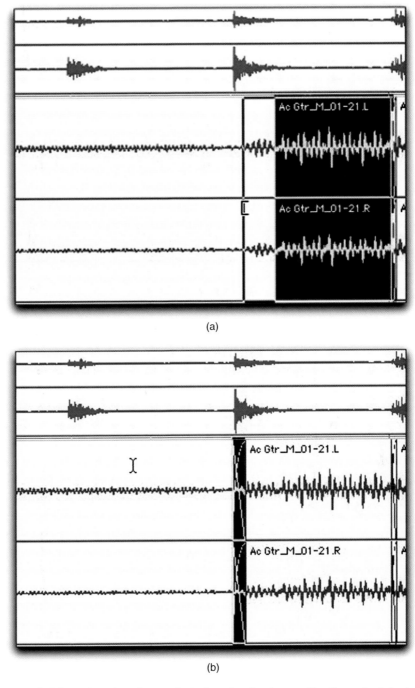

(a)

(b)

Figure 6.32 Trimming the pocketed region back over the time-stretched version and crossfading between them.

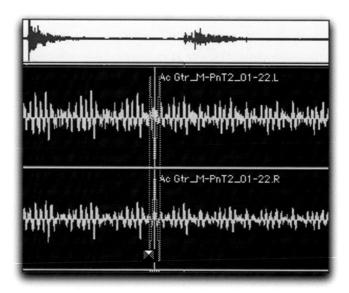

Figure 6.33 Crossfading between the original held note transient and its time-stretched later half.

Don't forget, any time you have a crossfade between a time-stretched region and its original, double-click the crossfade to bring up its dialog box, and change its shape to the standard linear in and out.

If you've had the loop muted, go ahead and turn it back on to listen to the acoustic against the loop. Even though we had a great guitar player, the performance should really be starting to feel nice and tight. Go ahead and pocket the rest of the acoustic guitar track against the hat loop initially, and against the drum tracks when they come in at the channel.

Acoustic pocketing summary

My standard approach to pocketing is going to change a little bit from song to song depending on a variety of factors. From what players you have in the session, to their performance quality, and even the style of music, every element of a session makes it unique to both the recording, mixing, mastering, and even the editing stage of the production. Issues to consider as you are making your own judgment calls on pocketing your records might relate to whether this is a Pop Country song with loops and percussion, that needs to be a little bit tighter. Or is it a 'friends in low places' track that can be a little looser to keep with the feeling of the song?

In general, your drums are going to be pocketed tighter than anything else. Your bass is going to follow with a pretty tight pocket right against the drums. After the bass usually comes the acoustic guitar. Starting with the acoustic guitar, you really need to let things breathe a bit more, without being quite as rigid in your pocketing. If you make the acoustic guitar as tight as you did your bass and drums, your track is going to start to feel mechanical, and every subsequent track will make the track feel increasingly stiff and quantized.

So, as you continue to work through this track, don't stress about getting those acoustic chords to the exact level of precision that you did on the percussion, drums, and bass parts. I don't mind leaving some notes a little behind, but you don't necessarily want any of the acoustic notes to be on top of the beat. There are a few exceptions of instruments that you will occasionally pocket on top of the beat, but that is usually some sort of lead instrument like an electric guitar solo or a fill line on a B3 organ. Any kind of line that's meant to be a stand-out line, and is supposed to be aggressive and more 'in your face', those are the sort of tracks that you can pocket a little on top of the beat. When it comes to anything else, like a strummed guitar in the background, we're going to let the drums lead the charge and pocket everything else behind them.

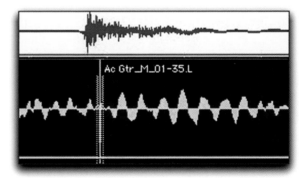

Figure 6.34   Pocketing the guitar track just a bit looser than the previous tracks.

Electricity in the air

Working with electric guitars

Alright. Now that we've made our way through the drums, bass, percussion, and acoustic guitars, let's take a look at some of our three tracks of electric guitar.

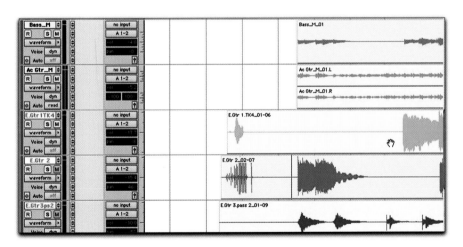

Figure 7.1 Our three tracks of electric guitar.

One at a time: Dealing with three different types of electric guitar

Go ahead and use the Show/Hide window to reveal our E.Gtr1, E.Gtr2, and E.Gtr3 tracks. If any of the tracks are inactive, select them and make them active. The first thing you'll notice visually about the electric guitars is that they are all doing very different parts. Unmute the first of the three tracks and let's take a brief listen to what it's doing.

The first note around bar 11 is a long sustaining strum. At least in this genre of singer/songwriter music, the nice thing about electric guitars is that they'll tend to be protracted and sustaining, rather than constantly changing and rhythmic like all of our previously pocketed tracks. Hence, they tend to be a bit easier and quicker to pocket.

Changing technique: The evolution away from transient pocketing

As we work farther and farther away from the drums, we're going to find ourselves less concerned with spotting a region or a note to an exact length of *xyz* milliseconds behind our guide track, and more concerned with whether the particular note feels right in the context of the mix and the track. Whether this is a heavy-handed piano performance, lush and vibrant strings and keyboard parts, or a soaring electric guitar, you're going to find more raked and drawn out transients, and less of what I call 'chunky parts'.

Cleaning up the intro on guitar 1

To start, let's just select and delete the unnecessary garbage at the beginning of the first electric part. We've talked before about just muting parts of the drums, but I know we're never going to use this piece of audio again, so let's just select and delete it.

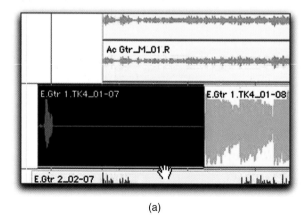

(a)

Figure 7.2 Deleting the noise at the top of the electric guitar track.

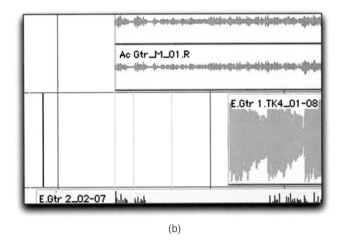

(b)

Figure 7.2   (Continued).

While we're at it, go ahead and clean up the intro entirely and add a tidy fade to the top of it so it doesn't get mucked up.

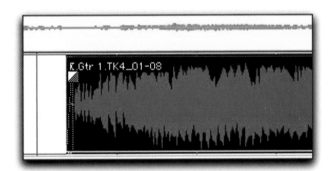

Figure 7.3
Cleaning up the
front of the electric
guitar.

Now that that's cleaned up, let's set up our screen with the tracks we want to see while pocketing. When working with the electrics, I like to actually leave the other guitars on the screen, as well as the obvious loop, kick, and snare. Let's go ahead and hide everything else, including the other electrics that we aren't working with right now.

Figure 7.4 The tracks to show and hide for pocketing electric guitars.

Dealing with dramatically mismatched volume levels

One problem we're going to encounter is the fact that the electric guitar was recorded *so much louder* than the other tracks, that when we zoom it out vertically to an acceptable level, the comparative waveforms of the bass and acoustic are very tiny. Que sera sera.

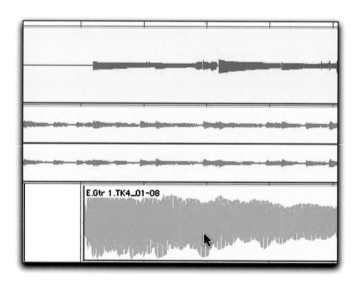

Figure 7.5 The electric guitar was recorded significantly louder than the other tracks.

Arranging your tracks, pocketing the electric

Finally, move your electric track up between your bass and acoustic guitar track to take advantage of their already pocketed waveforms.

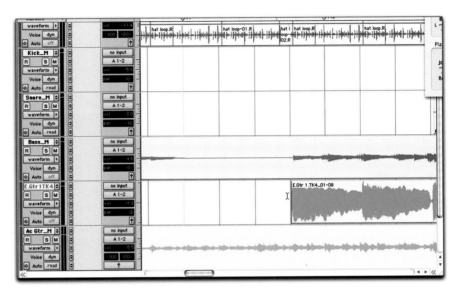

Figure 7.6   Moving your electric between the acoustic and bass tracks.

With the first note already sitting about where it needs to be, scroll down to the second note around 16|4|518. If you compare it visually and audibly against the downbeat performed by the bass guitar directly above it, you can tell this note is quite a bit early. Part of what appears to be an early transient on this specific note is actually the guitarist raking the strings. To get a better idea of what is going on with this second note, tab or scroll down to the end of it to see what the transition is between itself and the next note around bar 19. Here you'll see a pretty sizeable gap between the end of one strum and the beginning of the next. We'll see what we can do about that in a minute. For now, let's use our tried and true pocketing procedure, but with a special electric twist.

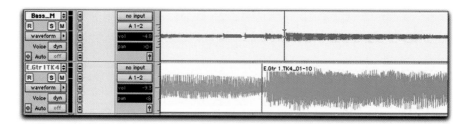

Figure 7.7 Relative to the bass guitar downbeat, the electric guitar chord seems a little early.

Nudge or drag the electric guitar over towards the right and pocket it just slightly ahead of the bass guitar note. Since this chord starts with the string raking sound we want it to lead the downbeat in.

Figure 7.8 Zoom out vertically and pocket the electric guitar slightly ahead of the bass transient.

The special guitar: Filling in gaps on a pocketed electric

Now here's where things get wacky. With a sustaining electric guitar chord like this, we can choose to fill in the created gap in one of two ways. The first we have covered repeatedly all the way from drums to bass to acoustic guitar. This would be our traditional time-stretching method, which, what the heck, we can cover at least once for the electric guitar.

1 Trim the end of the region to reveal the transient of the next note (Figure 7.9).
2 Insert an edit after the transient of the note we're about to time stretch (Figure 7.10).

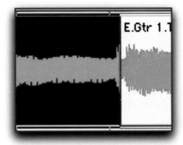

Figure 7.9

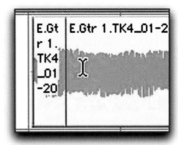

Figure 7.10

3 Use the TCE Trim tool to time stretch the later half of this note to over-lap the beginning of the next region (Figure 7.11).

4 Use the standard Trim tool to edit the next region's transient back into place (Figure 7.12).

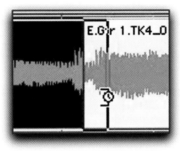

Figure 7.11

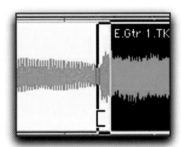

Figure 7.12

5 Crossfade over the out (Figure 7.13) and the in (Figure 7.14) points of the time-stretched region.

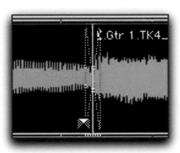

Figure 7.13

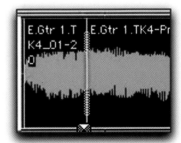

Figure 7.14

6 Double-click on the crossfades and change to a default linear crossfade shape.

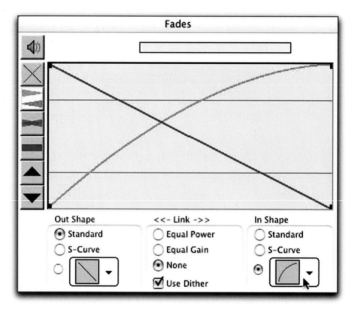

Figure 7.15   Change from this …

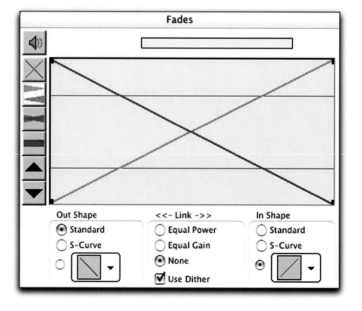

Figure 7.16   … to this.

Whew – OK, now here's the super secret method especially for electric guitars. Just drag the edit point back and crossfade, just like you've always wanted to do. That's it. If you want to take a walk on the wild side with me, undo through all of those edits you just made, until you're at the point where the pocketed second strum has a gap right before it. It will look something like Figure 7.17.

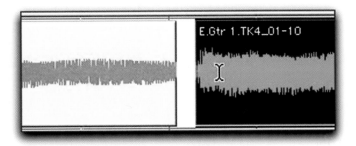

Figure 7.17   **Ready to take the shortcut.**

Are you ready? To fix this gap, just trim the right region to the left until the gap is filled and make a big old sloppy crossfade. Go wild man, go.

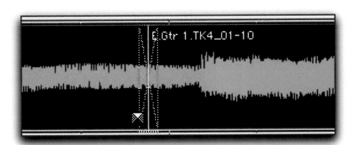

Figure 7.18   Blasphemy!

I know what you're saying: 'But you told us never to do that on a track! The audio before the crossfade is the same as the audio after the crossfade! Surely there will be some wicked bump or gurgling sound that will give me away for the sham of a Pro Tools editor I am'!

You are correct, I did say that. This brings us around to yet another time I get to make an exception to a previous statement (I'm thinking about running for public office). The fact is, on a distorted, hanging electric guitar part like this, you can get away with much more dramatic edits than you could ever get away with on the acoustic, piano, drums, or bass. As a matter of fact, we can get away with so much, you can literally create a huge wild crossfade like in Figure 7.19 to cover your edit – and probably get away scot-free.

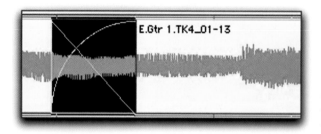

E.Gtr 1.TK4_01-13

Figure 7.19 Only on an electric guitar would an edit like this work.

Our friend masking, and his impact on electric guitars

At this point in editing the track, we're going to start making use of masking to cover simple things that even a golden ear could never hear in a million years. If you want to solo the electric guitar and listen to your crazy edit, you might be able to hear a slight difference as it rolls past. Bring in the rest of your tracks on top of it though – and unless you've done something insane, you'd never pick it out in a double-blind test.

So which method do you use? This is totally your call. You as the engineer/producer/editor know how much time, budget, and energy you have to work with. Equally obvious is that anything you do choose to do needs to sound great. Most of us are going to be dealing with budget issues for the majority of projects we're going to find ourselves in, and we'll likely not have hours and hours upon days and days and weeks to spend editing a given track. So, for the times when you are able to pull it off, and still have it sound amazing in the track, this can be a faster, cheaper, and just as effective way to do it, without having to utilize as much time stretching.

Spotting the electrics

Moving on to the next note, we have that gap we noticed earlier to deal with. The first thing I'm going to do is to add a cut right in that gap, then spot the later region to its original location.

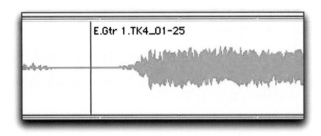

Figure 7.20 Cut in the gap around bar 19.

Figure 7.21 Looks like that second half is a little late – spot it back to its original location.

Spotting moves the region to the left about 200 ms. Since that both looks and feels good for that strum, use your Smart tool to trim the edges of the region a little closer to center, and add a crossfade.

Figure 7.22 Spotting moves the second region 200 ms left.

(a)   (b)

Figure 7.23   Trimming and fading the edit.

No more peaking: Letting your creativity flourish when editing electric guitars

Go ahead and let the track play through some of the first verse and listen to how it's working within the context of the song right now. Is it on top of the beat? Behind? A major issue we need to consider is how exposed is this track going to be in the mix? Currently we're listening to this track at around −7.4 dB, which is much louder than it will end up in the song. If you want to get a better feel for it, drop it down to around −13 dB and see if it bugs you when played back.

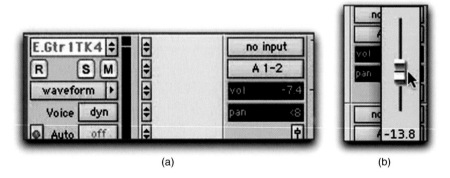

(a)   (b)

Figure 7.24   How loud will you have this track in the final mix?

Now we're dealing with tracks that will rely much more on your ear and creativity than on just the ability to define a clear transient and edit to a grid.

For example, check out the two strums that happen around bars 23 and 24.

Figure 7.25   This is two strums – can you tell the difference?

For me, for my taste, the note at 23 feels great, the one at 24 feels early. Looking closely, we might have a difficult time seeing where the transient lies for the second strum. At this point let's just make an edit around 24|1|346, and we'll then pocket the resulting region against the bass guitar note just above it and the acoustic guitar track below it.

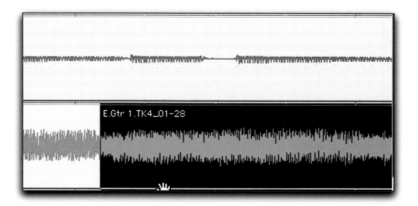

Figure 7.26   Separating out the early strum.

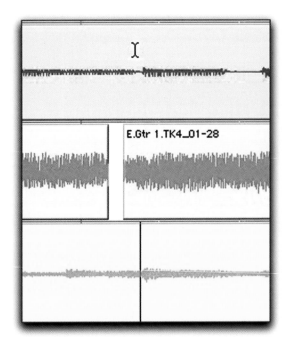

Figure 7.27 Notice the acoustic and bass are aligned. We've tried to align that dip in the electric transient against them.

Now, before we go to the effort of actually time stretching this region to fill in the gap, let's just drag back the later region to fill in the gap and cross-fade it. This way, if we pocketed incorrectly, we'll know it now before we spend all our time running the TCE gauntlet. Turn on your pre-roll again with Cmd-K (Mac)/Ctrl-K (PC), and listen back.

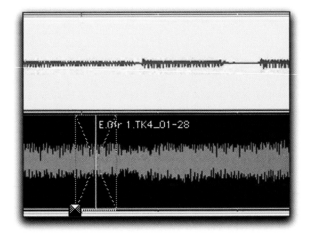

Figure 7.28 Checking the pocket first with a simple drag and cover edit.

Yikes. Looks like that waveform tricked us and we pocketed the audio way too late. In this case, what I usually do with an instrument that is not always easy to go into, look at, and effortlessly identify the transients is to either slow down the playback of Pro Tools (to hear the individual notes a bit better), or use a 10-millisecond nudge to bump the audio forwards and backwards in time until it feels right.

Identifying and fixing problems using slow playback and nudging

The first method just involves hitting Shift-Space bar to cause Pro Tools to play back the song at half-speed. This is also particularly effective if you need to give your talent a good scare in the middle of the session – say when they develop a bit too much diva-tude. Just hit Shift-Space bar and act shocked when PT plays their precious song back at a snail's pace. Firmly strike the side of the monitor as though to 'correct' the problem, and then play normally. Repeat this every time they get too picky about that 16th doubled vocal part. Say the system needs a rest and a reboot and go to lunch. Where was I again? Oh yeah.

The other approach, which I prefer, is to just select our misplaced region and use a 10 ms nudge to bump the audio forwards or backwards until it feels right. In this case, select it and nudge it earlier by about 40 ms. If you want, do a quick and dirty crossfade and play back to check it.

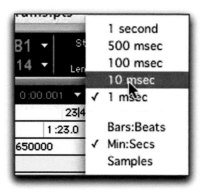

Figure 7.29 Setting a 10 ms nudge value.

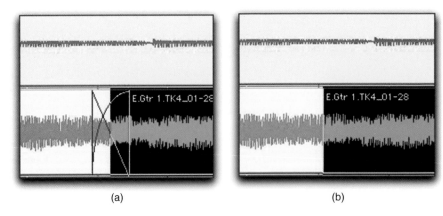

(a)   (b)

Figure 7.30 The region feels late. Hit your minus key three or four times to make the note happen 30–40 ms earlier.

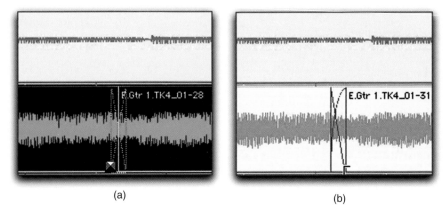

(a)   (b)

Figure 7.31 Using the old unscientific quick and dirty crossfade to check my pocket placement of this strum.

Booyah. Sounds great now. We were just off about 40 milliseconds, and the only way to really know it was just to nudge it back a bit and give it a listen. If yours still feels a little late or early, just nudge it back or forward a little more and try it again. Repeat until your track feels right. When you get it to where it feels good, drop in your crossfade and keep moving forwards.

Work your way through the rest of this electric guitar using the skills you've been developing all along the way. The most difficult part will be that you no longer get to just use a cookie-cutter, transient-grabbing approach once you get to tracks like this. Rather, every edit will develop your ear and test your creativity and ability. Initially, this will be a lot of trial and error as you develop your capacity to see the edit point in a constant waveform. The key is to always practice trusting yours ears even more than your eyes, and just keep nudging until it fits in the track. Utilize a swift drag and crossfade when it works, and the patient, more pure approach of the TCE method when it doesn't. Once you get to the end of the track, let's look at the second electric guitar for a different type of electric performance.

Electric guitar 2: Working with a percussive part

Like always, show your chosen track in the Show/Hide track list, and let's drag it up beneath the electric guitar 1 track, but above the acoustic track. Unmute it and give it a quick listen through to get a feel for what kind of performance is on this track. What you'll find is that this track has some percussive performance, as well as a fair amount of doubling parts. Interesting.

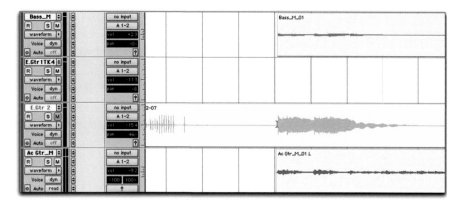

Figure 7.32   Your electric guitar 2 track in the mix.

Obviously, first things first. Select and delete the noise at the beginning of the track, add a tiny fade in, then let's listen to this first part that he plays.

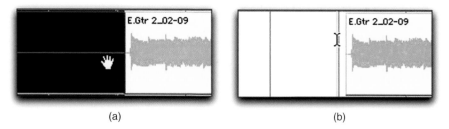

(a)   (b)

Figure 7.33   Deleting the noise at the top of the track.

The part itself sounds on, but I like to listen closely to fades, especially on subtle parts like this one. When you find the end of his fade, delete the noise between his fade and where he comes back in during V1B, and make a smooth fade out for his intro.

Figure 7.34   Unnecessary track noise.

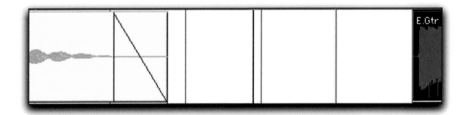

Figure 7.35   Much better.

Looking for pocketing guides: Parallels in other guitar tracks

Turn up this track to around −11 dB so we can focus on it a little better. When you listen back, it should become rapidly apparent that he is mirroring the part being played by the acoustic guitar. Since that is the case, we're going to vertically zoom in just a little bit and really scope it against the acoustic guitar. No longer are we just comparing a track to the kick drum or snare. Rather, we're really evaluating these notes and tracks in the context of our other, more rhythmic instruments that were pocketed against the kick and snare. In so doing we are eternally tightening up the performance, without ever making it sound mechanical. This is why we need to keep more and more instruments both in the mix and on the screen as we move up our pocketing ladder.

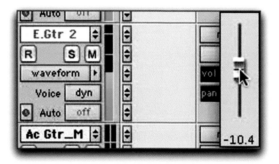

Figure 7.36 Turn it up.

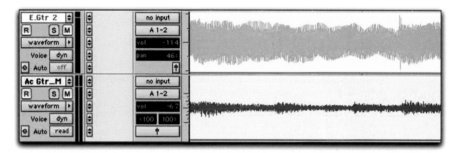

Figure 7.37 Since the second electric is mirroring the acoustic, we'll view them against one another, occasionally soloing just the two of them to focus.

So, since we know that he's mirroring the acoustic guitar part, you'll really be spending most of your time at the top of this track pocketing against the acoustic guitar waveform. For some great and clear practice notes, check out the waveforms around bar 16.

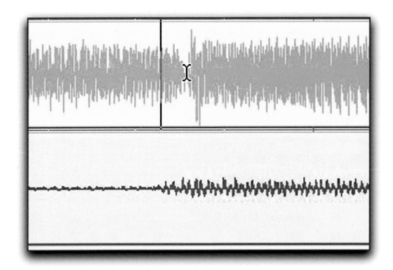

Figure 7.38   A good, clear practice note to pocket against the acoustic transient.

Since we already know how to work with those types of sounds, let's move down and find a special case around bar 40, right around verse 2. This transient heavy performance part takes us back to what we've already done with more of a straight-up pocketing approach, a la the drums. It's a chunky, rhythmic guitar part, without all the elongated notes. As a result, we can treat it just like we treated all of those other parts by dragging it up between the kick and snare temporarily for pocketing purposes. You'll probably also want the loop in view for good measure. In addition, solo it out for a minute to get a good idea of what the performer is playing.

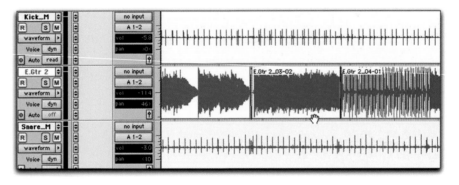

Figure 7.39   Check out the percussive electric part. Temporarily drag the track between the kick and snare.

A standard electric pocket

Let's turn the pocketing mill back on and work through the first couple of notes. Remember, this is one lead instrument that we don't have to worry about putting as far behind the beat as all the others. If it's a little bit on top we're probably OK. In addition, just like when we were looking at the longer sustaining electric parts, we really don't have to use as much time stretching to cover our edits when it comes to a low-level electric guitar in the mix. This should make it a moderately expeditious editing task to work through, as we will only need to use our TCE tool on any edits that don't sound very good with a simple drag and crossfade. So, starting with the first note around bar 40:

1 Pocket it directly behind the kick drum (Figure 7.40).
2 Trim the region to fill in the gap (Figure 7.41).
3 Drop in a quick crossfade to cover the edit (Figure 7.42).
4 Repeat with consecutive notes and regularly listen back (Figure 7.43). Sounds good!

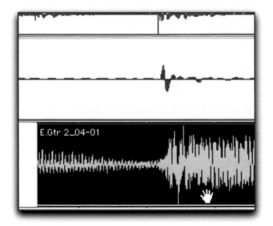

Figure 7.40

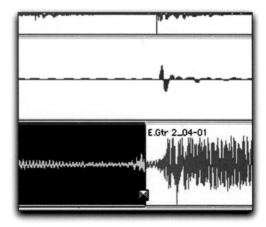

Figure 7.41

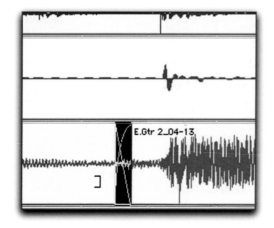
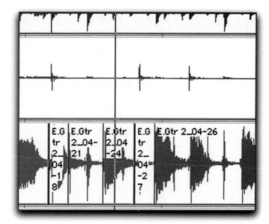

Figure 7.42   **Figure 7.43**

Some of the notes you will have to move by a pretty substantial amount. These sort of parts lend themselves to not quite being exactly on the beat. Another thing to note about these chunky distorted guitar notes is that the beginning of the note that you'll use to pocket is right before the loudest part of the transient. The waveform before that point is just fret or string noise.

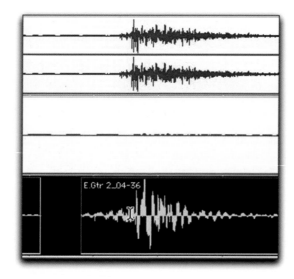

Figure 7.44   This note needed a pretty good nudge to get lined up in the pocket.

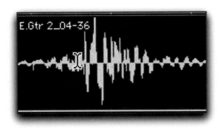

Figure 7.45 A closeup shows the beginning of the note used for pocketing.

A lack of time stretching and why

As you watch the accompanying DVD-ROM Movie files, you'll probably notice I do very little time stretching when editing parts like these. Personally, after years of trying a multiplicity of methods, I have found it to be an unnecessary step that I don't need to go through, and that doesn't improve the quality of my mixes. Could we get away with this kind of devil-may-care attitude when editing a piano track? Absolutely not. A percussive distorted electric guitar though? Yep. Just about every time.

Once you get done with this section and are back into the long sustained notes, drag your electric back down between your acoustic and electric guitar 1 tracks, and use them as your visual pocketing guide.

To wrap the electric guitars up, let's take a look at that third electric.

The third electric: The ambience track

Many times in a professional recording session a session player will lay down three different layers of electric guitar. The first two may be made up of rhythmic or percussive parts, but the third track will usually be made up of a more ambient sound. Such is the case with our third electric. Songs you may work on may have this sort of pad made from a keyboard track, or a soft synth, or even some small string part.

Begin by showing the third electric track and pulling it up below the first electric guitar part for now. Delete any excess noise at the beginning of the track and give the first section a listen.

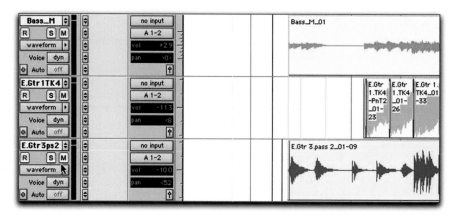

Figure 7.46 Bringing in our third electric guitar.

Visualizing the track: Finding places to pocket

You can basically tell just by looking at the long, sweeping, ambient wave-forms that this is not going to be difficult to pocket. Again, this sort of sound is generally best just separated and nudged into the place where it feels right, if it even needs to be nudged at all. A good exception is that third note that is actually a little two-note ditty that needs to be pocketed against the loop track (since the drums have not yet come in at this point). Drag the third electric track up to where it will be directly beneath the loop for the time being, and let's pocket that little double harmonic note.

Figure 7.47 This guy needs pocketing against the loop.

Figure 7.48 Move the E.Gtr3 track up under the loop for pocketing the harmonic double notes.

When you zoom in close you'll see that the first plucked note really sounds fine, but the second one needs just a bit of fine adjustment. Simply pocket it just a bit more behind the hat loop right above it and add your crossfade.

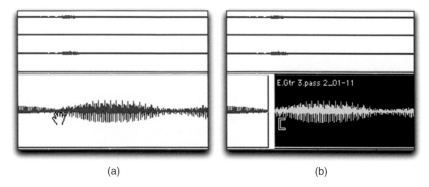

(a)   (b)

Figure 7.49 Pocketing the second harmonic note.

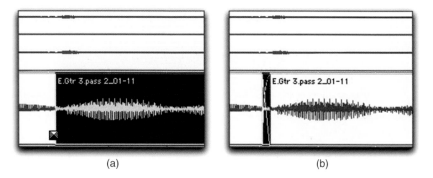

(a)   (b)

Figure 7.50 Trimming the edit back and making your crossfade.

There are several more of these little harmonic double notes that you can go in tight and really lock them to the groove of the song. For this type of track, you can really find yourself editing more aggressively than you do for the more base-level tracks. Simply listen closely, drop in an edit, slip them into time, crossfade, listen again, and move on. It really becomes pretty routine after you've done a few thousand edits. As they say, a journey of a thousand miles begins with a single edit point. Or something like that.

Spotting misadjusted guitar chords

In the case of this third track, after we get past the harmonic note section, the part changes back into big held strums around bar 19. Commonly,

257

when you've made a bunch of edits like you just did to the harmonic section, these big strums can look quite a way off. In this case, it looks very early compared to the loop track.

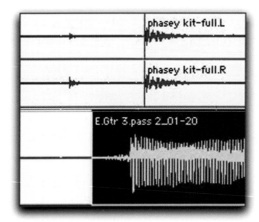

Figure 7.51 The strumming section looks really early.

This is no problem. Just jump over to Spot mode and click your region to spot it back to its original location. In this case it looks like it's about 40 milliseconds off. A click of the Original Time Stamp button and OK puts you back in business. Trim the gap out, crossfade, and move along.

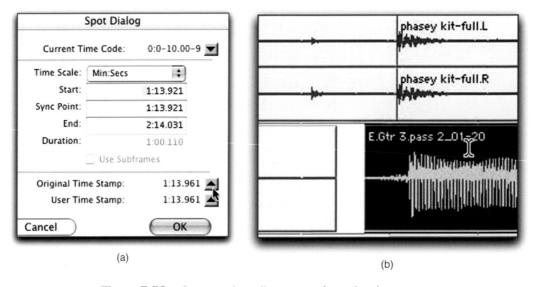

(a)

(b)

Figure 7.52 Spot mode pulls us out of another jam.

Pocketing against another electric

Once we get into this new section right before the channel, you will notice it's playing an almost identical part to the second electric guitar. So, when you go to pocket that section, drag the third guitar down next to the second guitar so you can use your (hopefully) already pocketed second guitar as a guide for pocketing this section of the third guitar track.

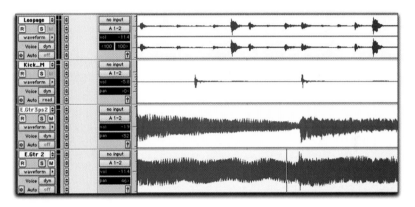

Figure 7.53 Pocketing the third electric against the second in the channel.

Electric guitar summary

So really, that's the multi platinum pocketing approach for bringing in all kinds of melodic instruments and parts. It's not as cut and dried as it was when we first started. It's not just about rhythm or just about the transients. It's really all about how these parts interact with one another, and how they interact with the vibe of the song. From section to section of the song we'll be listening to different things, soloing various tracks together, pocketing one part of a track against the drums, and another part against a guitar or bass. It's a continual ebb and flow. Make sure that you find yourself constantly listening to different things, and certainly not just listening and pocketing everything to the click. There's no quicker way to rob the life straight out of a song than to pocket everything in it to a metronome, leaving your tracks stale and not making sense with one another, or in the context of the whole.

Once you finish editing your guitar tracks, repeat the consolidation procedure and give them their own playlist, with the _M subscript in the track and file name for their master track designation. Now, we're ready to move on to vocals.

Autotuning: The not-so-dirty little secret behind a great vocal track

A brief discussion of tuning ethics

A lot of people assume that if you're pocketing an instrument or tuning a track, the musicians and singers must have not been up to snuff. At least for a lot of the major records, this couldn't be farther from the truth.

Take, for example, my editing on the 'N Sync record *Millennium*. Contrary to what some pundits may want you to believe, every one of the performers in that group can sing their hearts out. However even though they can sing just fine by themselves or in a group environment, consider for a minute just how much records have changed since the days when everyone 'had to be a great singer'. On many vocal-driven pop records, layers of background vocals can exceed 24, 32, or even 64 tracks of just vocals!!! Now for those of you with some singing experience, you can understand that even the most minor of pitch issues are going to become much more obvious and dramatic if you add that many layers. Now include the fact that vocalists will often sing their parts to a rough mix of all the other layers of vocals, as well as get their pitch cues from the other performances. Throw all that into a brew and even with the greatest singers on earth you are going to get tracks that will benefit from a musical ear, and the ability to hear and fix pitch.

Pocketing and tuning as a mix, rather than 'fix', issue

Pocketing and tuning are now less a matter of 'fixing' things than they are just a new part of the 'mixing' process. What these new technologies allow a great musician to do is focus their concentration on creating great parts and interesting performances, without worrying about 'Oh, that chord I hit was a little bit ahead', or 'Wow, that was a great feel, but it's just a few cents flat'. The same thing goes with tuning. Before these technologies, a final vocal performance from even the top Rock, Pop, and Country stars might be made up of dozens of punches done over hours and hours. The difference is that, now, rather than sitting in the control room, beating up our artist to get a performance with perfect timing, perfect inflection, perfect creativity, and perfect pitch, we can now focus on just getting a mind-blowing perform- ance, and easily fix the minor pitch issues that may come along with it.

This leads us into dealing with our current track. If you listen to the orig- inal performance you'll notice that our artist has both a pleasing and unique quality to his voice, and a good sense of pitch as well. He's really a lot of fun to listen to, and over the course of about five vocal takes was able to deliver us several great performances.

Starting with our comped track

There are a variety of popular methods for comping vocals, and really every producer has their own approach. One approach is to listen through your vocal takes in short phrases with your Solo set to X-Or in the Preferences. In addition, you may want to simply select over a small phrase and turn on the Loop Playback preference, then just go down the tracks one at a time listen- ing to each word, phrase, and sometimes even syllable for selecting, copying, and pasting down to your comp track. Since choosing the best performances from several wonderful options is far more 'art' than 'science', I feel the only effective way to show you is on the DVD ROM movie. So if you haven't already, go ahead and watch the section on comping and build your own vocal comp before you move on in the text. Alternatively, just choose your favorite among the included vocal takes, and proceed to Autotune with it. As always, for more information, as well as free videos, tips, and tricks, point your web browsers to www. multiplatinumprotools.com.

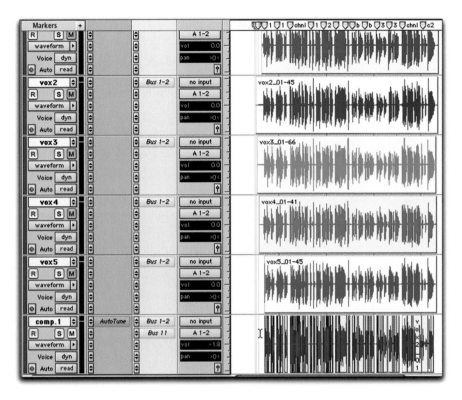

Figure 8.1 Your vocal takes with the vocal comp track at the bottom.

First things first: Getting to know the vocal

The first thing to do is make sure you are familiar with the vocal part. Since I've listened to this dozens of times when recording and comping it, I'm already ready to go. You should probably solo the comp track and give it a couple of listens to get a feel for his performance. Notice that the track really doesn't have any glaring pitch problems, but there are the standard subtle pitchy notes that we're going to just go in and nudge into place.

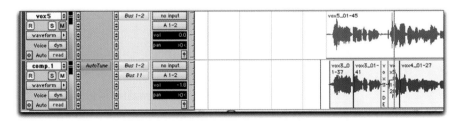

Figure 8.2 Solo and listen to the comped vocal track.

Notice that when I went in and cleaned up the track after comping, I created fades in and out of all the vocal phrases, deleted all the room noise, but was careful to leave the breath sounds in to keep the vocal sounding natural.

Setting up the Edit window for vocal tuning

To set up the screen to tune, we're going to hide literally everything except the comp track that we're going to be using.

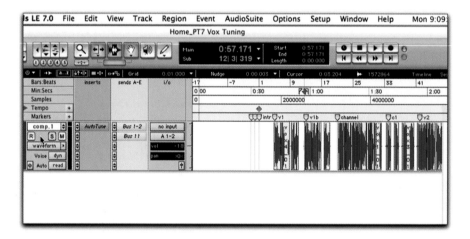

Figure 8.3 Hide everything but the comp track.

Go under the Track menu or use your keyboard shortcuts to add a single mono audio track beneath the comp track. Rename it as 'vox tuned'.

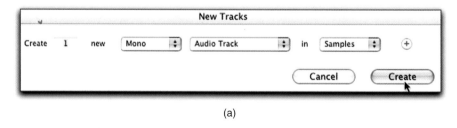

(a)

Figure 8.4 Create a new track and rename it as 'vox tuned'.

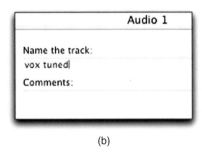

(b) **Figure 8.4**   (Continued).

Now that we have a track to record our tuned vocal to, we need to set up a Buss send to get our comped, tuned vocal track down to it. To begin, mute the comp track, and if there are any other bus send assignments on it (that may have been going to cues or FX during the tracking stage), go ahead and remove them by choosing 'no send' in their menu. Create a new send on the comp track that feeds any open and unused bus – let's use Bus 13 for this demonstration. After this, you can instantiate the AutoTune plug-in in the Inserts menu.

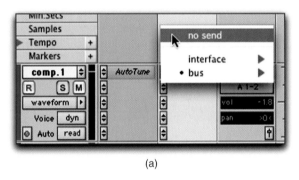

(a)

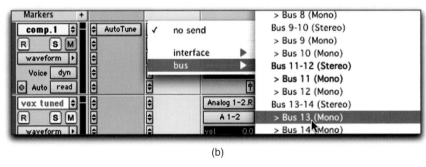

(b)

Figure 8.5   Remove any bus sends currently on the track, and create a single send going to Bus 13.

When the fader comes up for the new bus send, turn it up to unity with an Option click, and select pre-fader at the top. This will create a send from our comp track that is before its fader, but after the AutoTune plug-in.

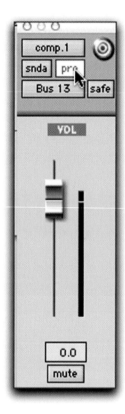

Figure 8.6 Turn your send up to unity and select pre-fader. Option~click (Mac)/Alt~click (PC).

Next, set the input of your vox tuned track to whatever bus you sent out of on the comp track – in this case, Bus 13.

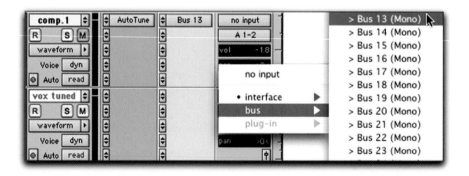

Figure 8.7 Set the input of your vox tuned track to Bus 13.

Under your Track menu, check to make sure Pro Tools is set to Input Only Monitoring. We usually run the rest of the recording session in Auto Input Monitoring, but for tuning purposes we need this set to Input Only Monitoring so that we're always listening to the input of the record-armed track and never the playback. Finally, make sure your vox tuned track is soloed and record armed and that your comp track is still muted. This will complete the process of setting up our routing.

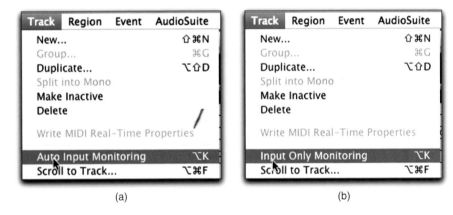

(a)   (b)

Figure 8.8 Make sure Input Only Monitoring is selected rather than Auto Input Monitoring. Option~K (Mac)/Alt~K (PC).

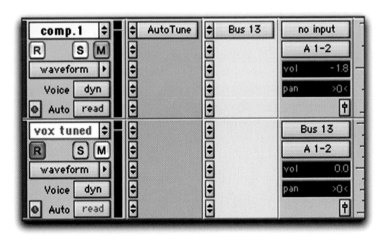

Figure 8.9 Your tuning setup should look something like this.

Where to start: Selecting your audio to be tuned

When tuning, I like to work with very small sections, usually no more than 6–8 seconds at a time. When you get too much material to cover, it just becomes too much to both listen to, as well as try to tune all in one setting in the AutoTune plug-in window. Go ahead and select the first couple of vocal phrases from about 41 to 49 seconds in the song.

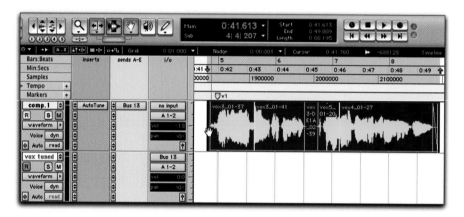

Figure 8.10   Selecting our first part to tune.

Click on the AutoTune plug-in on the comp track to launch the AutoTune main window, and let's take a look at the tools and features that we will be using for tuning this track.

Graphical versus Auto correction

First make sure you're in the Graphical editing window rather than the Autotuning window. The Autotuning window is the tired man's method of tuning that produces mediocre to mixed results when left to its own devices. It should never be trusted with a lead vocal, or really any type of lead instrument you may be needing to tune, especially if it's going to end up as a prominent track in the mix. It is the abuse of this window that has contributed to AutoTune occasionally being maligned in the press and by performers. The problem is not the Auto mode or its inherent quality. The real problem is when producers pass the tracks off to a recent recording industry graduate who has never been taught ear training or tuning skills and say 'Here, you know all that computer stuff – tune these vocals'.

Obviously the average person put in this situation is going to just set it to Auto mode and see what the plug-in comes up with. If they don't know the difference between a natural-sounding vocal and a techno bastardization – well, their traditional country record is going to suddenly have a harsh metallic element to the lead that will cause people to say 'What did they do to his voice?'

This is not to say the Auto correction mode is not useful. It is very useful and can be a real help when tuning backing vocals that are going to be down in the mix. However, we're never going to use it on a lead vocal. So, now that we're back from my tangent, just click over to Graphical mode.

Figure 8.11 The Graphical and Auto correction mode tabs.

Figure 8.12 The Autotuning correction window.

Notice that this AutoTune window has both a waveform already loaded into its memory (the red lines) and the pitch information ready to be edited (the yellow lines). We probably had the plug-in open before and took a brief look at the vocal part, but AutoTune doesn't automatically update to our currently selected audio every time we launch it. I know, it seems like an obvious feature need, but this is only the beginning of some quirky attributes of AutoTune – you'd better buckle in.

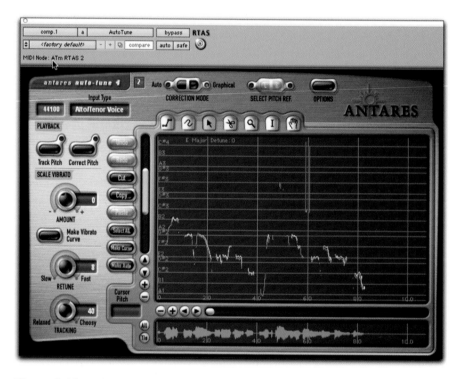

Figure 8.13   The Antares AutoTune plug-in.

Clearing out the old: Loading in the new pitch information

Before we can start the tuning process, we first need to select any of the old tracking data and delete it so we don't have to look at it. Just click the Select All button to select all of the yellow-lined pitch information (you'll notice little white editing points appear all over the yellow lines), and follow that up with the Cut button to toss it.

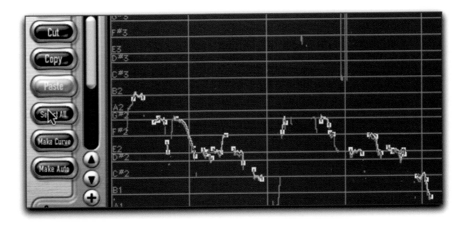

Figure 8.14 Selecting all of the old data. Hit the Cut button above to delete it.

Tuner settings: Getting the right Retune and Tracking speeds

Now we should check a few of our settings to get AutoTune set up the way we like it. When you first launch the plug-in, the Retune speed will be set to 20 and the Tracking speed will be set to 25 – in other words, a pretty fast and 'choosy' setting that will quickly retune most of the notes that it perceives as being off pitch. You are welcome to start there and adjust as you see fit, but when I'm tuning I like it to retune quite a bit faster, say down around 6 or 7, sometimes even faster. For tracking, a medium setting I've found that works pretty well for all different vocal styles is 40.

(a)

(b)

Figure 8.15 (a) The default settings. (b) My Tracking and Retune settings.

If you want to drag your plug-in window down a bit, it's always good to make sure you still have your selection highlighted and pre- and post-roll turned off.

271

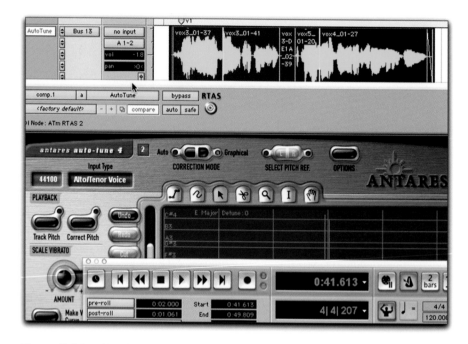

Figure 8.16   Making sure the regions are selected and pre/post-roll is off.

Now click the Track Pitch button (it will start blinking) and hit the Space bar to play back the selected piece of audio. This loads this audio and its associated pitch into AutoTune for our nefarious purposes.

Figure 8.17   Tracking the pitch is the first step to fixing it.

When it's done playing, you will notice AutoTune maps the audio in its grid-like window, with time on the horizontal axis and pitch on the vertical axis.

Viewing your pitches and setting your scale

Start off by using the Magnifying Glass to zoom in on the first couple of notes.

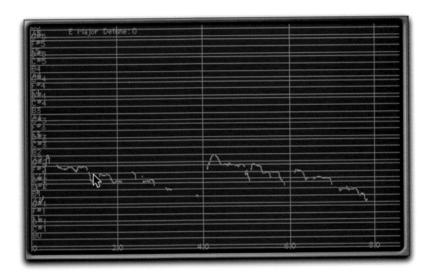

Figure 8.18 The grid showing the pitch over time.

Figure 8.19 The Magnifying Glass.

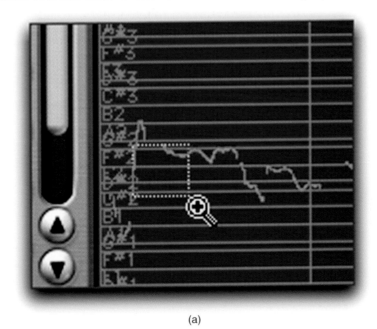

(a)

Figure 8.20 Zooming in on the early notes. The E major scale shows gridlines at every note in the major scale.

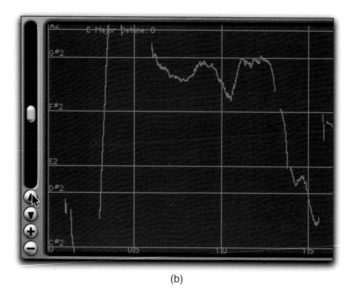

Figure 8.20
(Continued).

(b)

Now this song is in the key of E, so switch briefly to your Auto window and set your Key to E and Scale to Major. Then, when you switch back to the Graphical window the vertical axis will show a gridline at every note in the E major scale. If you know the performance has notes outside the scale, you can switch to Chromatic to help you find the closest note.

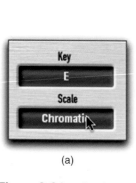

(a)

Figure 8.21 Setting the key and scale for the song. Chromatic will show every half note.

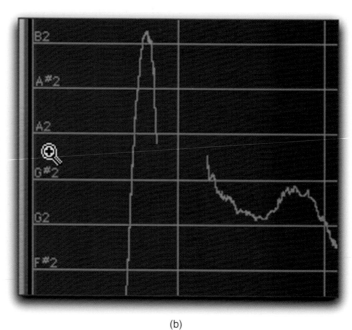

(b)

Using tools in the Grid window to correct the pitch

Grab the Insertion tool and drag over the first note in the Grid window. This will select the note that you wish to edit. Then, click your Make Curve button to make edit points appear at the ends of the selected pitch.

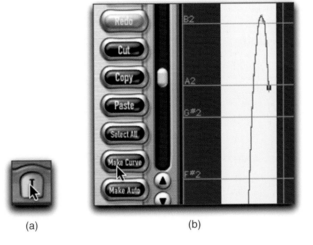

(a)   (b)

Figure 8.22
Choose the Insertion tool to drag select over the first note. Click the Make Curve button to add edit points to the selected pitch.

Change over to your Arrow tool and hover it over the selected pitch to turn your cursor into a crosshair. Once this is done, you will be able to click on and drag the pitch up or down to whatever note we desire.

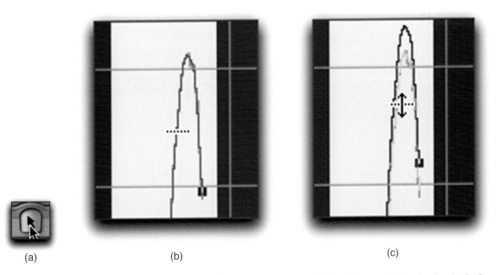

(a)   (b)   (c)

Figure 8.23   The Arrow tool lets us select our curve and drag it up to the desired pitch.

In this case, our artist was just touching the B pitch that he was going for, but I personally like to have the arc go a little above the line, sort of centered around it if you would.

That's it. You've tuned your first note. This was a subtle shift in pitch, but that's OK for your first time. Hit the Space bar to listen back to your first tuned note and make sure you didn't pitch it up too high or too low.

Now we're ready to start moving. Switch back to your Selector tool and let's repeat the process. Drag select over the next section and choose Make Curve to create your edit points. Switch back over to your Arrow tool and we're ready to adjust that note.

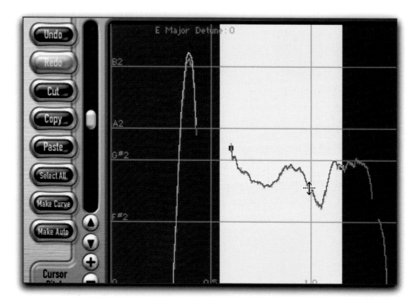

Figure 8.24   One note down, moving on to the second.

Using the Option key to lock in your pitch

Now, you may have noticed, if you're not very handy with a mouse, that when you grab the curve to adjust the pitch, it will let you drag the note both up and down, but also side to side!!! This can result in some pretty humorous pitch experiments, but nothing useful. So, especially for a really wide selection like we're looking at right now, if you hold down the

Option (Mac)/Alt (PC) key when clicking on the pitch to move it, AutoTune will allow you to drag the pitch up and down only, and not side to side, no matter how clumsy you are. Note the up and down arrows that appear over the crosshairs when you have the Option key pressed. If you hover your mouse over an edge, it will turn into a four-pointed arrow that will show you you can pull out, adjust, and stretch this pitch to virtually anywhere you like. It's pretty rare that you would use this, but there may be some surgical time where you have to get pretty crazy. Usually, though, I just use the Option key to make sure I don't accidentally move the note around in time. I've already pocketed the vocal after we comped it, so we're just looking to adjust the pitch right now.

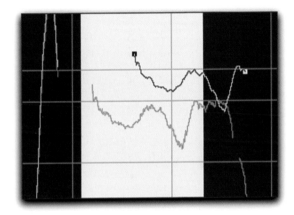

Figure 8.25 Oops! Forgot to hold down the Option key to prevent moving the note in time.

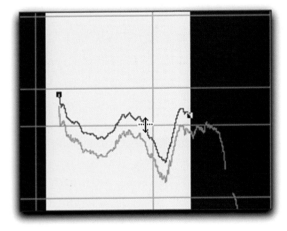

Figure 8.26 That's better. The up/down arrows tell us we can only move it in pitch, not time.

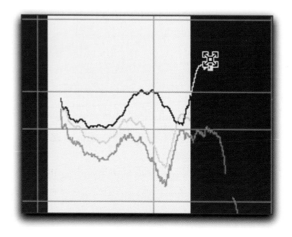

Figure 8.27 Hovering over an edge will let you drag the pitch and note anywhere you want to.

Alright, enough about possibilities, let's tune this note. Option-drag the whole section up until its middle bump is centered around the G#2. Then Option-drag the left edge of it down a bit closer to the G#2. Finally, recenter the entire note to where it's vacillating around the G#.

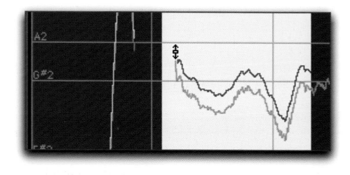

Figure 8.28 Centering the pitch, then Option-clicking the left edge.

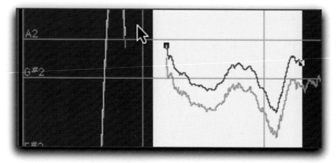

Figure 8.29 Where your pitch should end up – right around the G#.

Moving onto the later half of that note, use your Selector tool to select over it, Make Curve, switch to the Arrow tool, Option/Alt-click the pitch, and drag it up to where the front of the pitch is just over the G#.

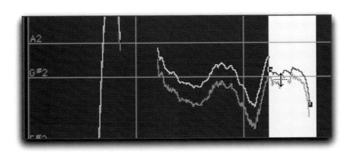

Figure 8.30
Tuning the later
half of our note.

Any time you want to de-select a piece of audio, simply click anywhere in the grid box outside of the selection. To select a pitch we've already edited, just click back on the yellow pitch curve you've adjusted.

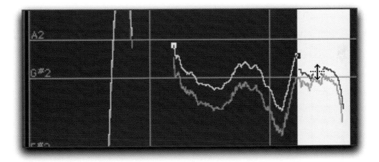

Figure 8.31   Reselecting the previously edited pitch.

Clicking to the left of, and dragging across, the Grid window will select multiple pitch curves and multiple edit points. To create additional edit points for a particularly difficult pitch, simply use the Scissor tool. The additional edit points will allow you to move individual pieces of the pitch curve independently of the others. See Figure 8.33C for an example.

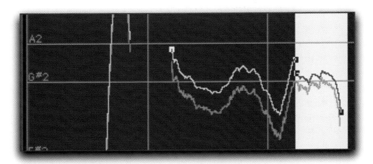

Figure 8.32   Notice both pitches are selected. Just drag across them to select multiple pitch curves.

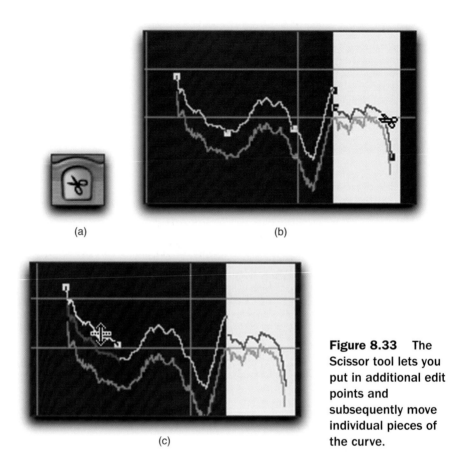

(a)   (b)

(c)

Figure 8.33 The Scissor tool lets you put in additional edit points and subsequently move individual pieces of the curve.

Undo doesn't work? What do I do to Undo?

If you're experimenting with these different tools as we go through them, you need to be aware that the standard Undo shortcut does not work within the AutoTune plug-in window. So if you chop a pitch up and move its pieces around for fun, then want to get back to where you started from, you have to use the AutoTune Undo button in the upper left corner of the window.

Figure 8.34 AutoTune has its own Undo button.

Use the horizontal and vertical scroll buttons to move along to the next note for selecting and tuning.

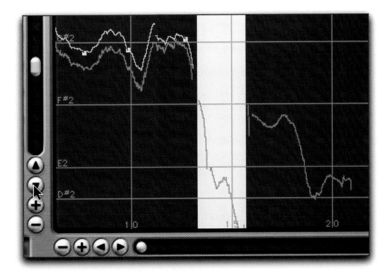

Figure 8.35 Use the scroll buttons to move over and view the next note.

Falloff notes and how to fix them

This falloff note goes by pretty quickly, making it a little difficult to know exactly what pitch to tune it to. Let's just tune the first half up towards the F# and center the second half over the E2.

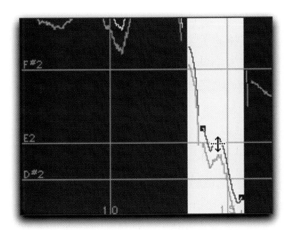

Figure 8.36 Tuning the falloff notes up to their closest pitch.

The next note moves from an F# pretty quickly down to an E, but he's just a few cents flat on both. The problem we encounter is that if we tune the first half up to the F#, the end of the note is now slightly above the E. No problem – use your Scissor tool to put an edit point right before that E and just tune that individual note down.

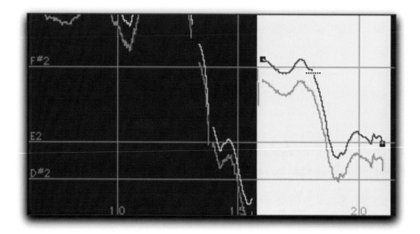

Figure 8.37   Tuning the first half of the note up to the F has made the later E note a little sharp.

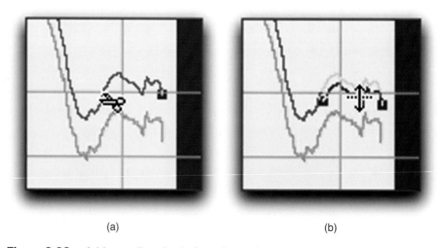

(a)   (b)

Figure 8.38   Add an edit point before the end note and center it around the E.

Listening back: Checking your work

Now that we've worked through the first few seconds of the song, it's a great place to pause and listen through what we've done so far. Just hit

Play and listen close to your perfectly pitched, but remarkably human, attempt at tuning. Nice job.

Since you're quickly becoming a tuning tour-de-force, take on the next note by yourself, but it should look something like Figure 8.39 around the F#2 when you're done.

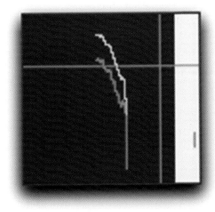

Figure 8.39 Just a quick minor tuning.

Now the next note after this is a bit of a special case, as the pitch waveform doesn't quite look like it's supposed to. It looks like the vocalist might have been climbing up to the pitch he was trying to hit, so I'm first going to select the left edge and center it to an even waveform, then tune the whole note around the E.

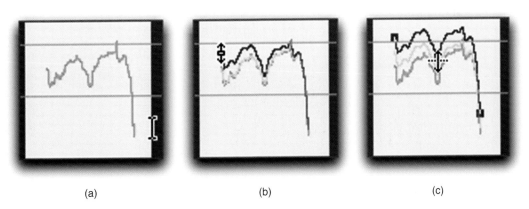

(a)   (b)   (c)

Figure 8.40 Centering the left side, then tuning the whole note to the E.

As we're working our way through this track, I should mention another reason (that's probably becoming apparent) that we just use small sections when tuning. Every time we want to listen back to our tune so far, we have to listen back from the very beginning of our selection. Even if we just want to listen to the last 2 seconds of a 12-second phrase, we have to listen to the first 10 seconds just to get there. So, it just makes more sense to do it in smaller sections.

Chopping up long notes

Just after the 4 grid line you'll notice a note that stretches out for some time. This may be a note that has a long vibrato or something to that effect. When you encounter a long note like that, don't be afraid to split it up into multiple edits and tune them individually. Often, on a really long note I may break it up into seven or eight different sections just to keep it sounding smooth. As long as the tuning makes them all sound good, they are good.

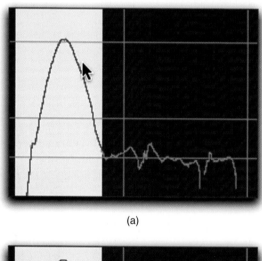

(a)

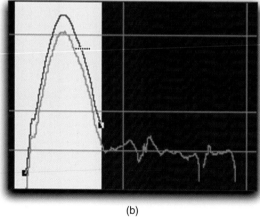

(b)

Figure 8.41 Selecting and tuning the first part of this note to just over the B.

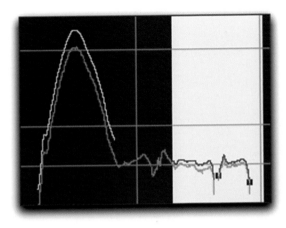

Figure 8.42 Tuning the last two pieces of the note individually.

On the next note between the 5|0 and the 5|5 in the AutoTune window, it looks like we need to tune the front, middle, and end separately. Start by tuning the end of the note with the right edge and an Option-click. Then select the middle of your note and use your Make Curve button to eliminate the curve you just made over the center section.

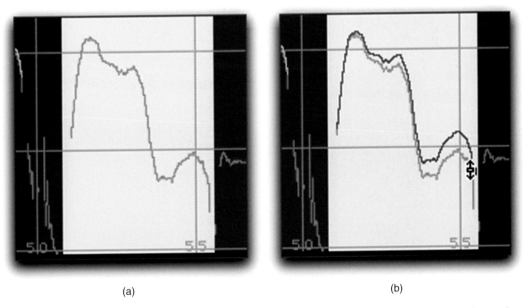

(a) (b)

Figure 8.43 This may need split into three parts for tuning. Start by tuning the end of the note.

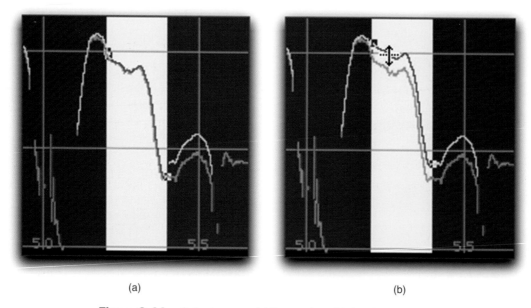

(a)   (b)

Figure 8.44   Select your middle section, Make Curve, then tune it where it needs to be centered around the G#2.

Whenever you need to make edits to multiple parts of a single note like this, I just like to always make sure there is overlap between my new curves. If I ever get my curves too far apart it may result in an audible bump, which obviously wouldn't be good.

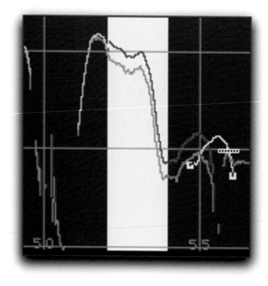

Figure 8.45   Don't move your pitch corrections within a note earlier or later in time without overlapping or it may result in an audible bump.

In defense of AutoTune: How other people do it wrong

Part of the reason I believe AutoTune is given a hard time for making people sound mechanical is because I see a lot of inexperienced tuners using the flat line drawing approach, and adjusting the Retune speed accordingly. While this obviously will get the singer dead on to the 'correct' pitch, it has the nasty side-effect of draining the spirit of their voice right out of the performance – leaving you with a cold metallic version of the original. It's just been my experience that when you use the Make Curve option, and try to keep as much of the original vocal inflection and timbre around the proper pitch, it creates a much less noticeable, but much more human-sounding, performance. This is just the method I use to make the vocal sound the least processed possible.

As you get closer to the end, if you listen back and hear anything that starts to bother you, just scroll back along to the part you need to tweak, grab your curve, and nudge it to where it needs to be.

For example, the word sung around 5|5 ends with a note that drops pretty far below the E, actually down past the D#. So, let's select right around the dropoff and see if we can't pull that note up closer to the E.

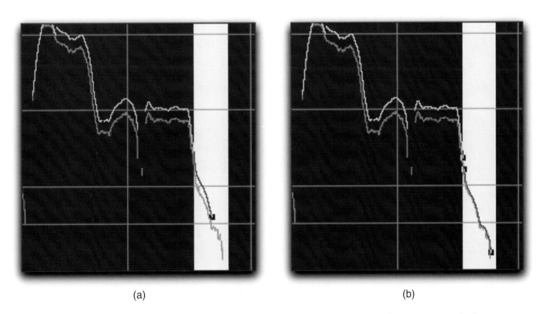

(a)   (b)

Figure 8.46   Select the dropoff. Make Curve eliminates our existing curve.

287

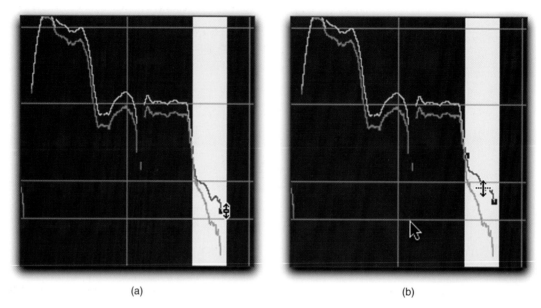

(a)   (b)

Figure 8.47   Select and tune the right side of the note (left), then tune the entire note to that E (right).

Continue to work your way through the other notes in the rest of this 8-second selection, and when you're done, we'll print it to our vox tuned track.

Printing the track

Once you're happy with the way the selection sounds, we're ready to print it. If you're feeling like it's just a bit too processed, even though you're happy with the graphs you've drawn in, you can always go in and slow down your Retune speed a little bit to, say, 18–20. The slower you go with that setting, the less you're going to hear the processing, for better or worse. On the other hand, if you're not hearing anything, and you want to keep that pitch pretty tight, keep your Retune speed up around 6–8.

Figure 8.48   Slow the Retune speed if you want to hear a little less processing and pitch correction.

To print it, simply hit Cmd~Space (Mac)/Ctrl~Space (PC) bar to record the tuned selection to the record-armed vox tuned track. A new, tuned, consolidated waveform will be recorded and given the filename 'vox tuned'. What could be better than that?

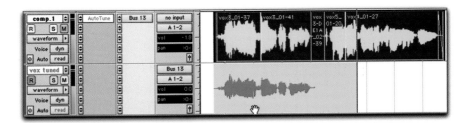

Figure 8.49   Recording/printing the tuned track.

From here through the rest of the track, it's just a matter of repeating this process. Select your next 8 seconds or so of vocal comp, and follow the remainder of the steps to tune the vocal.

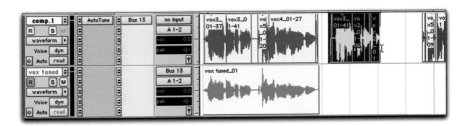

Figure 8.50   Selecting the next section – repeat the previous steps.

Tuning special cases and how to fix them

There are a few special cases I should point out as you work your way through the remainder of the vocal track. First, when I see a note that looks like a sine wave, and looks like it's on pitch, but sounds a little flat, I tend to nudge it up just a bit and split the difference with the waveform half above and half below the line.

Or take a case where the front half of a note is off pitch, but the back half is on. Don't be afraid to split the note in half and only tune the pitchy part.

Figure 8.51 Splitting the difference in pitch to set a note around its closest pitch.

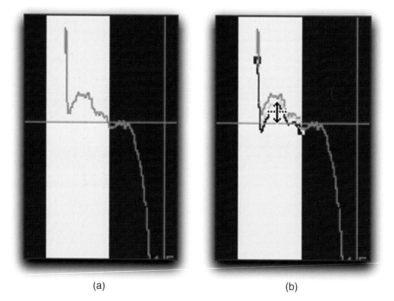

(a)   (b)

Figure 8.52 Tuning the first half of a note only.

The most important part

Above all, *listen often*!!! Sometimes you may miss a note. Or, if you happen to be a musician and know that you're in the key of E, even a region where you're not visibly sure what note they're supposed to be at, you can just compare it to the grid line and see 'Aha, they were trying to hit the E2'.

Otherwise, if you're not musical – well gosh. If you don't know anything about pitch you probably shouldn't be tuning a vocal.

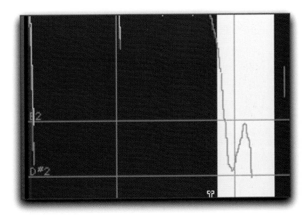

Figure 8.53   Which note is this supposed to be at? The D# or the E?

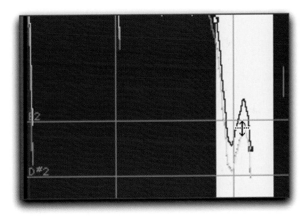

Figure 8.54   A musician knows to tune it to the E. Otherwise, try both.

'What about when there are several notes in rapid succession? Should I try to center each one of them around their proper pitch?' While this really depends on each case and the performance of those notes, in the series of rapid notes the singer hits in Figure 8.55, it already sounds good, so we're not even going to try to tune every note to an exact pitch. If you got a great performance to begin with, we only want to make it better, not microman-age it to death.

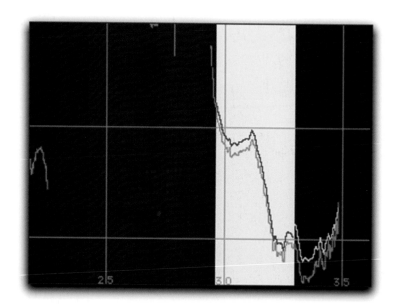

Figure 8.55 Fast falling notes. If it sounds good, don't try to tune them all to a pitch.

Like we've said before, unless you're tuning for an effect, the whole point is to get in and out without anyone being able to tell you've touched the track. That's why we go to the effort of using the Make Curve instead of just redrawing the waveform, or flatlining, or even autotuning them. It's a more musical way to do it, it's a more seamless way to do it, that's the way it's going to sound best, and make it the least obviously tuned.

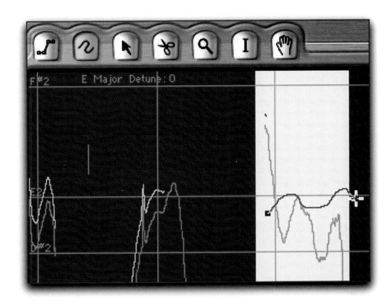

Figure 8.56 Using the Drawing tool to redraw the waveform.

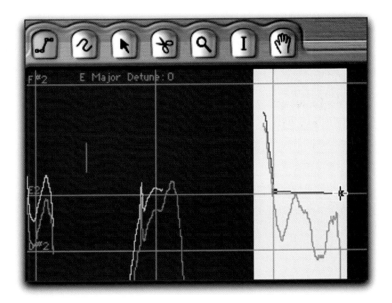

Figure 8.57 Using the Flatline tool to draw in a flat correct pitch will sound very unnatural with a metallic vibrato.

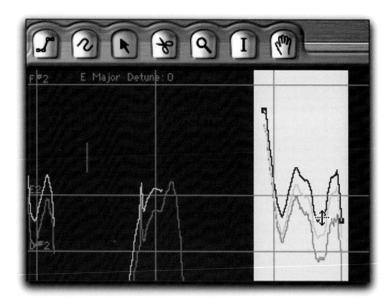

Figure 8.58   Using the Make Curve to create the correct pitch will sound very natural and smooth by keeping the shape of what was originally sung.

In the event that a pitch is just a little too unusual to be able to tune and make it sound natural, it would be better to just leave it alone and have it be a little flat or sharp than it would be to hear the tuning effect.

So that's basically the process for tuning any lead vocal or other important, prominent, and melodic mono track. The last thing we really need to look at for this session is a quick approach to tuning background vocals.

Tuning backing vocals

Working with backing vocals changes depending on the song and the nature of the background vocal in the song. There are times when I will graphically tune the backing vocals in a song just like I do the leads. In the case of a few of the backing vocals in this song, they are going to be so far in the blend of the mix we really don't need to spend that much time and effort pursuing twisted perfection.

Go ahead and show and solo the 'hi BG' and 'hi BG dbl' tracks. Listen to them to get a feel for what they're doing.

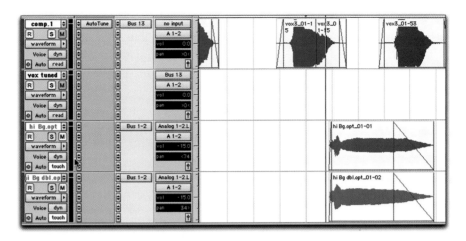

Figure 8.59   The backing vocal tracks.

The Chromatic scale and special tuner settings

It sounds like these are really just repeat lines, and since they are doubled they really don't have to be perfectly on pitch. So, let's actually use a tame version of the Auto mode in AutoTune to just even them out a little bit. Start by dropping AutoTune on the first backing vocal track, and set its input type to Alto/Tenor Voice, its Key to E, and its Scale to Chromatic. A lot of times if we try to set it to the major or minor key, AutoTune will try to snap it to the wrong note. The Chromatic scale will have AutoTune just pull the vocals to their nearest half note to what the singer actually sang. If your singer is a professional, as is the case in this track, Chromatic will be the better way to go, as a professional like this will never be off by that much.

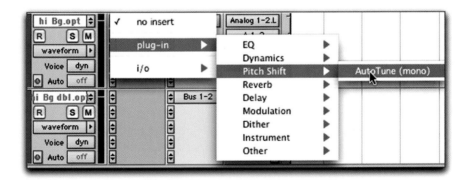

Figure 8.60   Put AutoTune on the first backing vocal track.

For your Retune speed, you won't want it really fast like we did on the lead vocal. I wouldn't do anything faster than a number in the forties personally. Forty is also a good number for your tracking when autotuning backgrounds.

Figure 8.61 Good settings for autotuning backing vocals.

Setting up your backing vocal tracks for tuning

Ungroup your backing vocals if they have groups turned on and we need to make a couple of tracks to print their tuned versions to. We could do this by just creating new tracks and resetting everything, but a quicker way to do it and keep all of our labeling similar from our old to our new tuned tracks is to just select the two tracks we want to duplicate, and go to the Track menu to choose Duplicate.

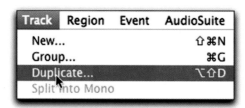

Figure 8.62 Duplicating our tracks. Cmd~D (Mac)/Ctrl~D (PC).

When the Duplicate Tracks dialog box comes up, turn off the Alternate Playlists and Group assignment options. We don't need these to carry over to our new tuned tracks.

Figure 8.63 Setting up duplicate tracks for tuning.

This creates two new tracks with all of their panning, inputs and outputs, and even volume automation already intact. Just delete the audio and remove AutoTune from the new track, set two fresh busses (say 15 and 16 this time) to unity, pre-fader, to go from our old tracks to the inputs of our new tracks, and you're ready to go.

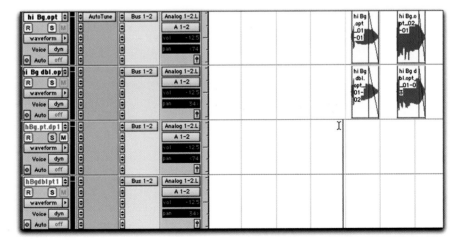

Figure 8.64 Setting up duplicate tracks for tuning.

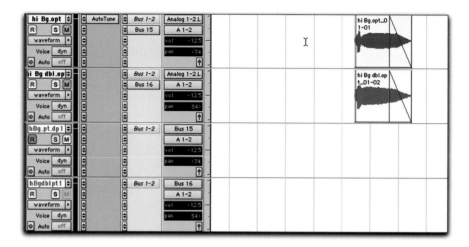

Figure 8.65 Your backing vocals tuning setup should look like this.

When you click on the AutoTune plug-in and play back the track you should see the red pitch bar bending right or left depending on the sharpness or flatness of the notes as they pass.

Figure 8.66 Notice the slightly sharp note. Not bad at all, but AutoTune will nudge it right back into place.

Listening back, this sounds pretty good, so I'm going to go ahead and print it just like we did with the lead vocal. If there's a little note or something that you're not sure about, you can always go back and fix it later. Once you've printed the first vocal, record arm the new double track, drag the AutoTune plug-in from the untuned backing vocals to the untuned backing vocals double, and repeat the process.

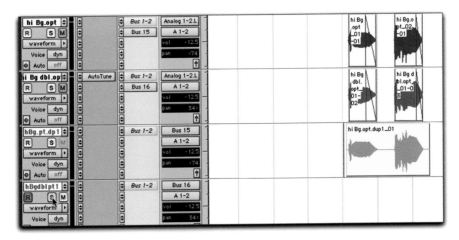

Figure 8.67 We've printed one backing vocal, now we're setting up the next.

Before we just print this one too, with a double track we don't want the Tracking at exactly the same speed, as that can often create a weird phasing effect. So what I do is often tune my first backing track a little tighter, but my double I will slow down the Tracking even more, say to 42, and my Retune speed down near 80.

Figure 8.68 AutoTune settings for the doubled backing vocals.

Listen through, and if that feels good, we go ahead and print it too.

Figure 8.69 Printing the second backing vocal with AutoTune in the foreground.

Solo both of the tracks and listen to them together now. I hear one tiny part where the pitch is not the same on the backing vocal and its double, so we can always drag AutoTune back up to the untuned backing vocal track and scope that one note a little closer in Graphical mode.

Remember, with background vocals, they are exactly that, in the background. We're going to be building a stack of vocals from the hi, mids, and low backing vocals and harmonies, and doubles, so we don't want every single note to be beating in time between them or it won't sound natural at all. The slight pitch and timing differences are what give it a choir sound, whereas if they are all the same the track will start to sound very mono, and very unison.

Finally, use this method to work your way through tuning the remainder of the backing vocal tracks to complete your edit of *Home*! Once you complete them, believe it or not, you're finished!!!

Wrapping it all up: A brief farewell

Now that you've worked your way through pocketing, editing, comping, and tuning all of the tracks in this song, bring up one of your own sessions while it's all fresh in your mind, and work your way through it to improve your skills.

From setting up your session and pocketing your drums, to time stretching those guitar notes and tuning your vocals, you'll be well on your way to becoming one of the 20 percenters we talked about earlier. So here's to you, one of the future Multi Platinum Pro Tools engineers.

Happy editing.

As always, for more free tips, tricks and videos, as well as to download Brady's final edited Pro Tools session of 'Home', point your browser over to www.multiplatinumprotools.com. To hear more from the tremendous artist you've just been working with, swing over and check out www.ericpaslay.com.

Index

80/20 rule, 4–5
Acoustic tracks, pocketing:
    about pocketing acoustic tracks, 207–8, 213–20, 231–2
    automation, turning off, 209
    chord transition identification, 213–14
    hiding most percussion tracks, 208
    no drum regions, pocketing techniques, 208–10
    pick/fingernail sounds, 217–18
    Pitch'n Time, correction with, 220–31
    raked chords, 211–13
    raking sound issues, 218–20
    setting up the acoustic pocketing Edit window, 208
Acoustic waveforms, 213–20
Activating tracks, 74–5
    Make Active command, 74
Antares AutoTune, 9–10
    see also Autotuning/AutoTune plug-in
Arrow tool, 275–6
Audio Media options, 133–5
Author, about the author, 1–2
Auto-Name Memory Locations When Playing, 17–18
Auto-Name Separated Regions, 18
Auto-scrolling disabling, 174
Automation, disengaging, 75, 209
Autotuning/AutoTune plug-in:
    about autotuning, 9–10, 287–9
    Arrow tool, 275–6
    checking pitch corrections, 282–4
    chopping up long notes, 284–6
    Drawing tool, 294
    Edit window setting, 264–7
    falloff notes, fixing, 281–2

flat sounding notes, 290–1
Flatline tool, 294
graphical versus auto correction, 268–70
half a note off pitch, correcting, 290–1
limitations/problems, 287–9
listening, importance of, 291–5
mechanical sounding issues, 287–9
as a mix rather than a fix, 262
monitoring settings, 267
old tracking data removal, 270–1
pitch correction with Grid window tools, 275–6
pitch locking with Option key, 276–80
pitch viewing, 272–4
printing the tuned track, 289–90
Retune speed settings, 271
scale setting, 272–4
Track Pitch button, 272
Tracking speed settings, 271
Undo problems, 280–1
vocal tuning settings, 264–7
see also Backing vocals, tuning

Backing vocals, tuning:
    about backing vocals, 295–6
    chromatic scale and special tuner settings, 296–7
    Duplicate Tracks, 297–8
    nudging with AutoTune, 299–300
    printing, 300–1
    setting up, 297–301
Barnett, Brady, 2, 6
Bass guitar pocketing:
    about bass guitar pocketing, 173–5, 205–6
    auto-scrolling disabling, 174

Bass guitar pocketing (*contd*)
   bumps/noise deletion, 189–90
   consistency, need for, 186–8
   Fade and Smart tools, 182–4
   finger/pick hitting string, fading out,
      180–4
   fret noise, 196, 199–200
   gap filling, 198, 200–5
   keying guitar to drums, 173
   note/waveform start, 180–1
   Nudge tool/menu, 183–4
   pocketing with Beat Detective, 184–5
   pre-rolls, post rolls and solos set up,
      188–9
   separating the note elements, 182–4
   Show/Hide bin for seeing transients,
      176–7
   Slip mode or Grid mode, 175–6
   Spot mode operations, 185
   transient matching avoiding
      dehumanization, 175
   transient removal, 192–4
   Zoom Waveform, using, 178–80
Batch Fade command, with Beat Detective,
   167–9
Beat Detective:
   about Beat Detective, 8, 137–8, 152–3
   with Analysis option to High Emphasis,
      142–4
   Batch Fade command, 167–9
   Beat Marker button, 144
   checking effects, 157–9
   clashes with master drum take, 159–67
   Conform button, 146–7
     conform issues, 154–7
   Copy and Paste instead, 170–2
   double transients, avoiding with
      Pre-splice option, 169–70
   Edit Smoothing, 148–51, 158
   Fill And Crossfade option, 149–51,
      158
   Fill Gaps option, 149
   frame of reference detection, 148
   launching, 140–2
   for lesser exposed instruments, 151–2
   Nudging for clash correction, 162–7
   Region Conformation button, 145–7
   Region Separation button, 144–5,
      153
   sensitivity/sensitivity bar, 142–3, 153

setting up, 138–40
   Trigger Pad option, 153–7
Brush track, pocketing, 136–7

Cher effect, 10
Chromatic scale and special tuner settings,
   296–7
Clip Indication option, 38
Comping, about comping, 8, 262
Conform, with Beat Detective, 146–7,
   154–7
Conversion Quality option, 24
Copy from source media option, 133
Copy and Paste:
   as alternative to Beat Detective, 170–2
   for gap filling, 203–5
Crossfade settings, 18–21, 23
   default settings, 21, 23
   with drum edits, 23
   and hard drive issues, 23–4
Crossfades, electric guitar, 240–2

Default Pro Tools settings, 14–15
Digi and Serato TC/E tools, 106–10, 226
Digital audio workstations (DAWs), 1, 4
Display tab:
   about Display tab, 38
   Clip Indication option, 38
   Default Track Color Coding, Peak Hold
      option, 38
   Draw Grids In Edit Widow option,
      38–9
   Draw Waveforms rectified option,
      39–40
   Edit/Mix Window Follows Bank
      Selection option, 38
   Flat List option, 39
   Organize Plug-In Menus By option, 39
   "Scroll to track" Banks Controllers, 38
   Tool tips Display option, 40
Dither, Use Dither option, 22
Double bass transients:
   cause, 191–2
   removal by nudging, 192–4
Double transients:
   with Beat Detective, 169–70
   with Pocketing, 102–4
Drag and cover edits, electric guitars,
   246–7
Drawing tool, 294

Drum editing/pocketing *see* Pocketing
Drum edits:
    crossfades with, 23
    fixing with TC/E, 104–6
Drums, editing within a drum group,
    82–3
DVD-ROM:
    contents, 10
    QuickTime 7 program, 11
    usage instructions, 11

Edit Insertion Follows Scrub/Shuttle
    option, 28–30
Edit window:
    cleaning up, 75–9
    Edit Window View selector, 43–4
    and Mix window, 42–5
    vocal tuning settings, 264–7
Edit/Mix Window Follows Bank Selection
    option, 38
Editing modes:
    about editing modes, 48–9
    Grid mode, 55–8
    Shuffle mode, 53–4
    see also Slip mode
    Editing within a drum group, 82–3
    Editing without crossfades, 195–6
Electric guitars:
    about electric guitars, 233, 259
    types of, 233–4
Electric guitars 1:
    cleaning up the intro, 234–6
    creativity suggestions, 244–7
    crossfade techniques, 240–2
    display techniques, 235–6
    drag and cover edits, 246–7
    gap filling, 238–42
    masking, 242
    pocketing, 237–42
    separating strums, 245
    slow playback and nudge technique,
        247–9
    spotting, 242–4
    TC/E trim tool usage, 239
    track arranging, 237–8
    transient pocketing, reduced use of, 234
    volume level mismatch issues, 236
Electric guitars 2, percussive part:
    about percussive performances, 249–50
    pocketing guides, 250–2

pocketing techniques, 253–5
    setting up, 249–50
    time stretching, lack of, 255
Electric guitars 3, ambience track:
    about the ambience track, 255
    finding pocketing places, 256–7
    pocketing against the second guitar,
        259
    spotting/correcting misadjusted chords,
        257–8
Enable Session File Autobackup option,
    34–5

Fade, default settings, 18–21
Fade length, 22–3
Fade shape settings, 18–21
Fade tool, with Bass guitar pocketing,
    182–4
Falloff notes, fixing, 281–2
Fingernail/pick sound problems, 180–1,
    217–18
Flat List option, 39
Flatline tool, 294
Fret noise, 196, 199–200

Gap filling:
    about gap filling, 197–9
    with Copy and Paste, 203–5
    using time compression/expansion
        (TC/E), 200–3
Grabber tool, 49–51
Graphical and Auto correction, 268–70
Grid mode, 55–8
    with bass guitar pocketing, 175–6
Guide track finding, 79–81
    loop tracks, 79–80
    shaker transients, 79–81

Hard drive issues, with crossfades, 23–4
Hat tracks, 84–5
Hiding tracks, 76–8
History, of performance creation, 4
Hole filling *see* Gap filling

Import Session Data window, 132
Importing tracks, 71–4
    importing session data, 71–2
    percussion tracks, 132, 134–5
    Show/Hide window/tracks box, 73–4
Inactivating tracks, 76–8

index

Keyboard Focus mode option, 32
Keyboard shortcuts, 65–6
Kick and snare tracks, 82–4
    working without, 88

Latch Record Enable Buttons, 30
Levels of Undo option, 25–6
Link options, 21
    Link Mix And Edit Group Enables, 30–2
    Link Record/Play Faders, 32–4
    Link Timeline and Edit Selection, 45–6
    Link to source media, 133
Listening, importance of, 291–5
Loop tracks, 70

Masking, electric guitars, 242
Matching Start Time options, 25
Mechanical sounding issues, 287–9
Mix and Edit windows, 42–5
Monitoring settings, autotuning, 267

Noise cleanup, 88–92
    see also Gap filling
Noise from transients:
    cause of, 191–2
    removal by nudging, 192–4
Nudge tool/menu:
    bass guitar pocketing, 183–4
    with gap filling, 198
Nudge value options, Beat Detective, 162–7

Open Ended Record Allocation, 35
Opening a session, 69–71
Operation tab:
    Edit Insertion Follows Scrub/Shuttle, 28–30
    Timeline Insertion Follows Playback, 26–7, 30
Option (Mac)/Alt (PC) key, 45
    for pitch locking, 276–80
Organize Plug-In Menus By option, 39

Percussion edit window, 135
Percussion tracks, importing, 132, 134–5
Pick/fingernail sound problems, 180–1, 217–18
Pitch:
    checking corrections, 282–4
    correction with Grid window tools, 275–6
    half a note off pitch, correcting, 290–1
    locking with Option key, 276–80
    Track Pitch button, 272
    viewing, 272–4
    see also Autotuning/AutoTune plug-in
Pitch'n Time tool:
    acoustic track correction, 220–31
    Digi's default TC/E plug-in, 226
    Serato, 106–8, 226
    see also Autotuning/AutoTune plug-in
Pocketing:
    about pocketing, 6–8, 92–6
    adjusting bounds to fit fades, 113–16
    brush track, 136–7
    counter changing, 94
    crossfades with, 98, 113–16
    drum pocketing system, 96–9
        checking with pre-roll, 116–17
        leaving feel, 122–3
        saving pocketed drums, 128
        wrapping up the edit, 124–30
    as a mix rather than fix, 262
    purpose of, 7
    snare hit, 95–6
    Spot mode rescue, 119–21
    Tab to Transient feature, 121
    transient choosing, 92–3
    transient timing, 95
    transients doubled, 102–4
    transients missing, 99–102
    trim tool usage, 97–8
    for vocalists, 7
    when to leave alone, 117–18
    see also Acoustic tracks, pocketing; Bass guitar pocketing; Electric guitars; Time compression/expansion TC/E
Pre-splice option to avoid double transients, 169–70
Preferences/Pro Tools Preference box:
    about Pro Tools option, 15–16
    Auto-Name Memory Locations When Playing preference, 17–18
    Auto-Name Separated Regions, 18
    Conversion Quality, 24
    Default Fade Settings, 18–21
    Edit window and Mix window, 42–5
    Edit Window View selector, 43–4
    Enable Session File Autobackup, 34–5
    fade length, 22–3

306

Fade shape (Fade In/Crossfade/Fade Out) settings, 18–21, 23

Keyboard Focus mode option, 32

Latch Record Enable Buttons, 30

Levels of Undo option, 25–6

Link Mix And Edit Group Enables, 30–2

Link options, 21

Link Record and Play Faders, 32–4

Link Timeline and Edit Selection, 45–6

Matching Start Time, 25

Mix window and Edit window, 42–5

Open Ended Record Allocation, 35

Options (Mac)/Alt (PC) key, 45

Regions List, 46

Show/hide button, 46–8

Solo Latch, 36–7

time-stretching, 21–2

Tracks List, 46–7

use All Available Space, 35

Use Dither, 22

see also Display tab; Operation tab; Processing tab

Printing the tuned track, 289–90

backing vocals, 300–1

Pro Tools:

basic principles, 3

books about, 5

history of, 4

see also Preferences/Pro Tools Preference box

Processing tab, TC/E (Time Compression/ Expansion), 40–1

QuickTime 7 program, 11

Raked chords, 211–13

Raking sounds, 218–20

Regions List, 46

"Scroll to track" Banks Controllers, 38

Scrub tool, 28

Selector tool, 52–3

Serato and Digi TC/E tools, 106–10, 226

Session opening, 69–71

Setting up Pro Tools preferences, 13–14

default settings, 14–15

settings choosing, 14–15

Setup menu, 16

Shaker transients, 79–81

Shortcuts, keyboard, 65–6

Show/hide button, 46–8

Shuffle mode, 53–4

Slip Mode, 49–53

with bass guitar pocketing, 175–6

grabber tool for grabbing/moving regions, 49–51

Selector tool, 52–3

Trimmer tool, 54–5, 56

Slow playback and nudge technique, electric guitars, 247–9

Smart tool, 66–7

with Bass guitar pocketing, 182–4

Snare hit, 95–6

Solo Latch option, 36–7

Spot mode, 119–21

with bass guitar pocketing, 185

Spotting, electric guitars, 242–4

Stylus RMX, 71

Tab to Transient feature, 121

TC/E *see* Time compression/expansion TC/E

Techniques, how to acquire, 2–3

Time compression/expansion TC/E, 9

about time-stretching, 21–2

with acoustic tracks, 220–31

with electric guitars, 239

for fixing a drum edit, 104–6, 110–13

with gap filling, 200–3

Serato and Digi TC/E tools, 106–10, 226

TC/E tool option, 40–1, 104–6

transients retention, 110–11

Timeline Insertion Follows Playback option, 26–7, 30

Tool tips Display option, 40

Track menu:

disengaging automation, 75

Make Active command, 74

Track operations:

activating tracks, 74–5

excess noise cleanup, 88–92

grouping tracks, 86–7

guide track finding, 79–81

hat tracks, 84–5

hiding tracks, 76–8

importing tracks, 71–4

inactivating tracks, 76–8

kick and snare tracks, 82–4

track naming, 83–7

working without kick or snare tracks, 88

Transient matching avoiding dehumanization, 175

Transient retention for natural effects, 110–11

Trigger Pad option, with Beat Detective, 153–7

Trim tool, 54–5, 56, 57, 97–8, 198

Tuning:
 about tuning/autotuning, 9–10
 Antares AutoTune, 9–10
 Cher effect, 10
 ethics of, 261
 graphical versus auto correction, 268–70
 as a mix rather than fix, 262
 see also Autotuning/AutoTune plug-in

Undo:
 Levels of Undo option, 25–6

Undo problems, 280–1

Use All Available Space option, 35

Use Dither option, 22

Vocals/vocalists:
 about vocals, 263–4
 backing vocals, 295–6
 pocketing, 7
 see also Autotuning/AutoTune plug-in

Zooming/Zoomer tool, 58–65
 horizontal zooming, 58, 60
 with pocketing, 99–101
 Vertical Zoom, 59, 60
 Zoom Toggle function, 61–3
 Zoom Waveform, using, 178–80